Principles of Organic Farming: Textbook

(As per Syllabus of 5th Dean's Committee, ICAR)

Principles of Organic Farming: Textbook

(As per Syllabus of 5th Dean's Committee, ICAR)

P. L. Maliwal

ICAR-Emeritus Professor

MPUAT, Udaipur-313001

(Former Director Research, MPUAT, Udaipur)

SCIENTIFIC
PUBLISHERS

Published by:

SCIENTIFIC PUBLISHERS (INDIA)
5A, New Pali Road, P.O. Box 91
Jodhpur 342 001, India
E-mail: info@scientificpub.com
Website: www.scientificpubonline.com

Print: 2025

ISBN: 978-93-89184-49-5
eISBN: 978-93-89184-51-8

Printed in India

भारतीय कृषि अनुसंधान परिषद
कृषि अनुसंधान भवन-II, पूसा, नई दिल्ली 110 012
INDIAN COUNCIL OF AGRICULTURAL RESEARCH
KRISHI ANUSANDHAN BHAVAN-II, PUSA, NEW DELHI 110 012

डा. नरेन्द्र सिंह राठौड़
उप महानिदेशक (कृषि शिक्षा)
Dr. Narendra Singh Rathore
Deputy Director General (Agril. Edn.)

Phone : 011-25841760 (O)
Fax : 011-25843932
E-mail : ddgedn@gmail.com
nsrdsr@gmail.com
Website: www.icar.org.in

June 4, 2019

FOREWORD

Increasing awareness about environmental pollution, deteoriating soil productivity and health hazards associated with indiscriminate use of fertilizers and pesticides and consumers preference for safe and quality food are the major factors that lead to growing interest in alternate forms of agriculture in the world. Organic farming is the production system gaining importance world over as it is eco-friendly and safe to human health. The demand for organic food is steadily increasing both in developed and developing countries with an annual average growth rate of 20-25%.

Organic farming is a holistic system designed to optimize and fitness of diverse communities within the agro-ecosystem, including soil microorganisms, plants, livestock and people. Presently organic farming is being practiced in 181 countries with the organic area of 69.8 million hectares. In India the area under organic farming is 3.429 million hectares, which includes certified cultivated area and wild certified area during 2018-19. The registered organic farmers in the country are highest in the world. The growth of organic farming is around 12-15% in India.

The author presented the book entitled "Principles of Organic Farming: Textbook" which is a scientific and systematic collection of text in a cogent and easily understandable manner. The book is written to fulfill the basic requirement of under graduate students of agriculture stream considering the syllabus recommended by 5th Dean's Committee of 1CAR.

I hope that this book will serve as text book for under graduate students of agriculture. I understand that this will also be of equal significance for university teachers and extension personnel's dealing in organic farming. I congratulate Dr. P. L. Maliwal for his painstaking efforts in bringing out this book covering the latest innovations in the field of organic farming under the ICAR- Emeritus Professor Scheme.

I am confident that this book will be widely accepted among the students. I extend my best wishes to Dr. Maliwal for success of this book.

(N.S. Rathore)

Preface

Organic agriculture is defined worldwide as *"farming without the addition of artificial chemicals"*. Organic farming is widely practiced in under developed and developing countries, largely because of the fact that the farmers are unable to afford costly inorganic inputs. However, in developed countries people are concern about environmental safety and sustainability of land productivity and health hazards, which encouraged organic farming in these countries. The strategies evolved during green revolution era cannot be valid any more under the prevailing conditions. A new strategy of living with nature and nurturing it for sustainable high productivity should be evolved and organic agriculture showed the way to effectively use the available natural resources for the benefit of mankind without polluting the environment.

A large number of books are available on organic farming but most of them are much elaborated and it is very difficult for the students, teachers to screen the material as per their requirement. Therefore, efforts have been made to compile and consolidate the information available on organic farming and presented in a cogent, clear and simple language. The book is written basically to fulfill the requirement of under graduate students of agriculture stream considering the syllabus of 5th Dean's Committee of ICAR. The author is optimistic that the students community will harness the benefit of this book in understanding the subject of organic farming.

The author is deeply indebted to Indian Council of Agricultural Research, New Delhi for providing opportunity for preparing the manuscript under the Emeritus Professor Scheme of ICAR.

The author acknowledge his indebtness to authors of books and other publications from which most of the material/literature have been drawn. In several cases, it has not been possible to obtain permission for reproduction, for which the authors and publishers offer their apologies.

The author is thankful to Dr. A. Joshi, Dean, RCA, Udaipur and Dr. Dilip Singh, Professor and Head, Deptt. of Agronomy for their generous support in bringing out this publication.

I am also thankful to Dr S.K. Sharma, Zonal Director Research, Dr S.L. Mundra, Director Extension Education, MPUAT, Udaipur, and other faculty members of Department of Agronomy, Rajasthan College of Agriculture, Udaipur for the help and suggestions rendered for betterment of this book.

The author also welcome constructive comments and suggestions from the readers whosoever for further improvement of the book and that would be duly acknowledged. I sincerely thank to the Scientific Publishers, Jodhpur for publishing this book well in time.

P.L. MALIWAL

Contents

Chapter

1

An Introduction to Organic Farming

In the view of increasing consumption of fertilizers and insecticides, deteriorating soil health and pollution of air, water, and food, there is a growing concern about global environment and so the concept of organic farming is gaining importance world over in order to develop sustainable and eco-friendly agricultural production system. Organic farming is primarily based on the principles of use of natural onfarm organic inputs like farmyard manure (FYM), compost, green manures, oil cakes, press mud, etc. besides the natural biological pest control and plant protection measures with minimal use of natural minerals to promote agro-economic system and soil biological activity.

Prior to independence, Indian agriculture was considered to be a gamble of monsoon. There was great uncertainty in food grain production as it was solely dependent on quantum of rainfall and its distribution. In India in 1950^{s} and 1960^{s} the country faced severe food scarcity and govt. of India was forced to import food grains from other countries. Looking to this peculiar situation in the country, India aimed to boost the food grain production drastically. Under the stalwart leadership of Dr. M.S. Swaminathan the green revolution programme was launched by the Government of India in 1960^{s}, in which major emphasis was given on improved crop varieties and use of fertilizers and pesticides. Indian farmers adopted these practices and shifted from organic based farming to chemical based inorganic farming, so called modern agricultural.

No doubt this has revolutionized agriculture in the country and led to substantial increase in food grain production in few years and the country has become self sufficient in this respect. But latter it has been realized that the increase in food grain production was achieved at the cost of deteriorated soil health, environmental pollution and health hazards, which costed us dearly. This has led to think about organic farming, which required total elimination of the fertilizers and pesticides/ chemicals in agriculture. It is high time to rethink that we should switch on to chemical less agriculture or maintain a balance between the two to get required food grain production vis-a-vis protect the soil, environment and health of the people, which is of foremost importance in future. The aim of the organic farming is to protect the environment as well as natural resources (land, water and plants) and produce safe and healthy food.

Organic farming is a holistic system designed to optimize the productivity and fitness of diverse communities within the agro-ecosystem, including soil organisms, plants, livestock and people. The principal goal of organic production is to develop enterprises that are sustainable and harmonious with the environment. Organic farming promotes the use of crop rotations and cover crops, and encourages balanced host/predator relationships. Organic residues and nutrients produced on the farm are recycled back to the soil. This system merits consideration on the ground that most of the ill effects of modern day agriculture are avoided. Use of agrochemical is forbidden. There is emphasis on build up of organic matter in the soil, thereby activating biological activity. Soil is treated as a living organism. Maintenance of favourable soil structure and use of crop rotation that improves soil fertility, control insect pests, and diseases, and weeds through organic means only and get nutritious healthy food.

The modern concept of organic farming combines the tradition, innovation and science. Although, history states that the movement for organic way of life recognized in 1905, it could gain ground after realizing the ill effects of modern agriculture in the late 1990's. In 1905, the British botanist Sir Albert Howard often referred to as the father of modern organic agriculture, documents traditional Indian farming practices, and came to regard them as superior to conventional agriculture science. During 1940, in Japan, Masanobu Fukuoka, a microbiologist working in soil science and plant pathology, quit the job as a research scientist, returned to his family's farm, and devoted the next 30 yr to develop a radical no-till organic method for growing grain, known as "Fukuoka farming". Many other practices such as *Rishi krishi,* Natueco farming, *homa* farming, *panchagavya krishi* and biodynamic farming are associated with organic agriculture.

1.1 DEFINITION

Organic farming is defined differently by different organizations as follows:

United States Department of Agriculture (USDA) study team on organic farming defined organic agriculture as "organic farming is a system which avoids or largely excludes the use of synthetic inputs (such as fertilizers, pesticides, growth regulators, livestock feed additives, etc.) to the maximum extent feasible, rely upon crop rotations, crop residues, animal manures, legumes, green manure, off-farm organic wastes, mechanical cultivation, mineral bearing rocks, and aspects of biological pest control to maintain soil productivity and tilth, to supply plant nutrients, and to control insects, weeds, and other pests. It considers that soil is a living system which must be "fed" in a way that does not restrict the activities of beneficial organisms necessary for recycling of nutrients and producing humus at centre to this concept.

According to the **Codex Alimentarius Commission** the definition says "organic agriculture is a holistic production management system which promotes and enhances agro-ecosystem health, including biodiversity, biological cycles, and soil biological activity. It emphasizes the use of management practices in preference to the use of off-farm inputs, taking into account that regional conditions require locally adapted systems. This is accomplished by using, where possible, agronomic, biological, and mechanical methods, as opposed to using synthetic materials, to fulfill any specific function within the system." To promote organic agriculture and to ensure fair practices in international trade of organic food, the Codex Alimentarius Commission, also framed certain guidelines for production, processing, labelling and marketing of organically produced foods.

According to the **International Federation for Organic Agricultural Movements (IFOAM)** definition "organic agriculture is an agriculture system that promotes environmentally, socially, and economically sound production of food, fiber, timber, etc. In this system, soil fertility is seen as the key to successful production. Working with the natural properties of plants, animals, and the landscape, organic farmers aim to optimize quality in all aspects of agriculture and the environment."

1.2 CONCEPT

Organic farming endorses the concept that the soil, plant, animals and human beings are linked. Therefore, its goal is to create an integrated, environmentally sound, safe and economically sustainable agricultural production system. Soil is a living system linked with different natural components. Human interact with these natural components (minerals, organic matter, micro-organisms, animals and plants) to achieve harmony with nature and create a sustainable agricultural production. A key feature of organic farming is the primary dependence on natural resources and those developed locally (green manures, crop residues, farm wastes, etc.), rather than external inputs (especially synthetics). The farmer manages self-regulating ecological and biological processes for sustainable and economic production of products. Organic farming systems do not use toxic agrochemical inputs (pesticides, fungicides, herbicides and fertilizers). Instead, they are based on development of biological diversity and the maintenance and replenishment of soil productivity.

Thus in today's terminology it is a method of farming system which primarily aims at cultivating the land and raising crops in such a way, as to keep the soil alive and in good health by use of organic wastes (crop, animal and farm wastes, aquatic wastes) and other biological materials along with beneficial microbes (bio-fertilizers) to release nutrients to crops for increased sustainable production in an eco-friendly pollution free environment.

Organic farming describes two major aspects of alternative agriculture:

Substitution of manures, farm organic resources and bio-fertilizers (INM) for inorganic fertilizers.

Biological and cultural management of insect pests, diseases and weeds (IPM, IDM and IWM) instead of chemical control.

1.3 PRINCIPLES

In view of indiscriminate fertilizer and pesticide use, deterioating soil health and pollution of air, water, and food there is growing concern about global environment and so the concept of organic farming is gaining growing importance world over in order to develop sustainable and eco-friendly agricultural production system. Organic farming is primarily based on certain principles. The fundamental rules on which principles of organic farming work are as follows:

(i) The principles are to be used as a whole. They are composed as ethical principles to inspire action.

(ii) These principles are the roots from which organic agriculture grows and develops. They express the contribution that organic agriculture can make to the world and a vision to improve all agriculture in a global context.

(iii) Agriculture is one of humankind's most basic activities because all people need to nourish themselves daily. History, culture and community values are embedded in agriculture.

(iv) The principles apply to agriculture in the broadest sense, including the way people tend soils, water, plants and animals in order to produce, prepare and distribute food and other goods.

(v) They concern the way people interact with living landscapes, relate to one another and shape the legacy of future generations.

(vi) The principles show that organic farming is much more than the renunciation of agro-chemicals.

1.3.1 The Principle of Health

The principle of health aims that organic agriculture should sustain and enhance the health of soil, plant, animal, human and planet as one and indivisible. This principle points out that the health of individuals and communities cannot be separated from the health of ecosystems - healthy soils produce healthy crops that foster the health of animals and people. Health is the wholeness and integrity of living systems. It is not simply the absence of illness, but the maintenance of physical, mental, social and ecological well-being. Immunity, resilience and regeneration are key characteristics of health. The role of organic agriculture, whether in farming, processing, distribution, or consumption, is to sustain and enhance the health of ecosystems and organisms from the smallest in the soil to human beings. In particular, organic agriculture is intended to produce high quality, nutritious food that contributes to preventive health care and well-being. In view of this it should avoid the use of fertilizers, pesticides, animal drugs and food additives that may have adverse health effects.

1.3.2 The Principle of Ecology

The Principle of ecology aims that organic agriculture should be based on living ecological systems and cycles, work with them, emulate them and help to sustain them. This principle roots organic agriculture within living ecological systems. It states that production is to be based on ecological processes, and recycling. Nourishment and well-being are achieved through the ecology of the specific production environment. For example, in the case of crops this is the living soil; for animals it is the farm ecosystem; for fish and marine organisms, the aquatic environment. Organic farming, pastoral and wild harvest systems should fit the cycles and ecological balances in nature. These cycles are universal but their operation is site-specific. Organic management must be adapted to local conditions, ecology, culture and scale. Inputs should be reduced by reuse, recycling and efficient management of materials and energy in order to maintain and improve environmental quality and conserve resources. Organic agriculture should attain ecological balance through the design of farming systems, establishment of habitats and maintenance of genetic and agricultural diversity. Those who produce, process, trade, or consume organic products should protect and benefit the common environment including landscapes, climate, habitats, biodiversity, air and water.

1.3.3 The Principle of Fairness

The principle of fairness aims that organic agriculture should build on relationships that ensure fairness with regard to the common environment and life opportunities. Fairness is characterized by equity, respect, justice and stewardship of the shared world, both among people and in their relations to other living beings. This principle emphasizes that those involved in organic agriculture should conduct human relationships in a manner that ensures fairness at all levels and to all parties - farmers, workers, processors, distributors, traders and consumers. Organic agriculture should provide everyone involved with a good quality of life, and contribute to food sovereignty and reduction of poverty. It aims to produce a sufficient supply of good quality food and other products. This principle insists that animals should be provided with the conditions and opportunities of life that accord with their physiology, natural behavior and well being. Natural and environmental resources that are used for production and consumption should be managed in a way that is socially and ecologically just and should be held in trust for future generations. Fairness requires systems of production, distribution and trade that are open and equitable and account for real environmental and social costs.

1.3.4 The Principle of Care

The principle of care aims that organic agriculture should be managed in a precautionary and responsible manner to protect the health and well being of current and future generations and the environment. Organic agriculture is a living and dynamic system that responds to internal and external demands and conditions. Practitioners of organic agriculture can enhance efficiency and increase productivity, but this should not be at the risk of jeopardizing health and well being. Consequently, new technologies need to be assessed and existing methods reviewed. Given the incomplete understanding of ecosystems and agriculture, care must be taken.

This principle states that precaution and responsibility are the key concerns in management, development and technology choices in organic agriculture. Science is necessary to ensure that organic agriculture is healthy, safe and ecologically sound. However, scientific knowledge alone is not sufficient. Practical experience, accumulated wisdom and traditional and indigenous knowledge offer valid solutions, tested by time. Organic agriculture should prevent significant risks by adopting appropriate technologies and rejecting unpredictable ones, such as genetic engineering. Decisions should reflect the values and needs of all who might be affected, through transparent and participatory processes.

1.4 SCOPE

The organic farming is not a technology by itself. But it consists of set of practices evolved around several principles based on widely varying character or resources in different ecosystems. The totality of each ecosystem and mechanism of delicate balance available between components of each ecosystem would ultimately decide which principle would weigh more or less and ultimately which practice should be predominant in organic farming.

In this background, the assessment of scope of organic farming should adopt widely varying techniques for different ecosystems and for different principles of organic farming. Further, the assessment should also include the harmony achieved between new set of organic farming practices and components of ecosystem.

The scope of organic farming cannot be assessed on an uniform scale at global level. The criteria of assessment and in turn the extent of scope of organic farming vary according to soil, climate, cropping system, vegetation, irrigation, allied activities (animal husbandry, sheep rearing, poultry, piggery, fisheries, etc.) as well as harmony achieved between different components of each eco-system.

Scope of adopting organic farming exists in every farming situation with varying degree. Within the meaning of broad concepts and varying principles of organic farming, it can be adopted for every crop and in every situation. Each farming situation offers unique scope of practices that may be specific to such situation. But a clear analysis of all the features and components of each situation should be undertaken to identify constraints and resources to evolve such practices. Some situations (dry ecosystems) may vividly offer lesser scope for adopting the principles on *in-situ* decomposition. But, they would be ideal situations for using pre-decomposed manure. It is only required to plan, evolve and implement the practices in the right perspective.

As compared to chemical farming, the organic farming has no much scope, if dramatic results are expected in a short time. In some situations, the organic farming practices may not be cost-effective in short run as compared to use of fertilizers which are supplied at subsidized rates. However, the assessment of scope of organic farming on such consideration is indeed undermining the serious damages caused by chemical farming to our most ecosystem. The need of the hour is not to relegate the importance of organic farming by such considerations but to regenerate and strengthen the self-supportive production system, which was cheaply damaged in a short span of 30 to 40 yr. Several scientific studies have been reported to show that organic farming system can increase the crop yield and stabilize it to the level of chemical farming in long run, although in the initial 1 to 2 yr, yields may be reduced to some extent. The long term perspective of practicing organic farming can lead to a better scope in gain from a cropping system. Organic farming does not only aim at higher crop yield or returns, but, it aims at developing long term self and sustainable practices which not only result into increase crop yield/ returns but also sustain the whole production system to achieve perpetual productivity. In this perspective, the organic farming has excellent scope and should be encouraged not for short term gains but for long term perspective.

Although the scope of adopting organic farming exists in every farming situation, the evolution and adoption of practices involves a great deal of planning and physical work on the part of farmer as compared to chemical farming. Simple and short-cut solution such as use of fertilizers to supply plant nutrients is certainly an easy job as compared to making use of all available organic sources in a farm to convert them into available plant nutrients. Similarly, chemical control of a pests or diseases may involve a great deal of understanding the pest dynamics and its place *vis-a-vis* other biological and abiotic components in a given ecosystem. But to control it by organic farming techniques, it needs greater planning and understanding the life cycle of pest, its food habits, its breeding habits and its place in an ecosystem.

Chemical farming can never be a compatible component in any agro-ecosystem and it has to be followed as an extraneous component forced upon an ecosystem. Application of any manufactured agrochemical would not match with the natural processes and such chemicals serve as impending factors for many natural processes of an ecosystem. In organic farming, the practices are so developed that they are compatible with most of natural processes of the ecosystem and thus accepted by nature without any impairment of the harmony evolved by an ecosystem. For these reasons, organic farming practices are regarded as self-supportive and sustainable in the long run. The scope of adopting organic farming practices should not be curtailed for the reasons of some difficulties in evolving and adopting them. The scope of organic farming has to be perceived in the long run vision of developing self supporting sustainable production system than short run minor impediments.

1.5 ORGANIC V/S NATURAL FARMING

There is a misconception that organic farming is merely to say "no" to chemicalism. But apart from restricting and to the extent possible eliminating chemicals (pesticides & fertilisers), it has something else also to convey. One who understands the whole concept of organic farming will be certainly inspired by it. The differences between organic farming and natural farming (based on natural principles) are given below:

Natural farming	Organic farming
(i) It is not alternative system of farming but part of the philosophy of life involving continuous search to know the true spirit and form of nature	(i) In many respects close to natural farming – but does not have the philosophical over-tone of natural farming

Natural farming	Organic farming
(ii) Totally eliminates all the components of modern farming	(ii) Organic farming does not totally exclude elements of modern farming. It involves limited and essential farm operations and organic nutrients and pest management
(iii) It indicates a 'do nothing' approach	(iii) It indicates a soil building programme – more intensive style of natural farming. Application of natural plant protection chemicals (which are not inorganic derivatives), use of organic manures (instead of chemical fertilizers) are permitted.
(iv) The essential principles are: • No cultivation • No chemical fertilisers • No weeding • No plant protection	(iv) Principal elements are: • Maintaining a living soil through minimum tillage • Making available all essential nutrients through organic means • Weed control with non-chemical alternatives • Biological pest control

1.6 SCENARIO OF ORGANIC FARMING

Presently organic farming is being practiced in 181 countries of the world. The ill-effects' of chemicals used in agriculture have changed the mindset of some consumers in different countries who are now prefer to buy organic food even with high premium for health. Policy makers are also promoting organic farming for restoration of soil health and generation of rural economy apart from making efforts for creating better environment. The global organic area is 69.8 mha roughly along with 61 standards and 468 certification bodies. The growth of organic farmland and organic share during 1999-2017 is depicted in fig. 1.1. The important countries with large organic areas as given in fig. 1.2 includes Australia (27.15 mha), Argentina (3.01 mha) USA (2.03 mha), Spain (2.02 mha), China (2.28 mha), Italy (1.80 mha), France (1.54 mha), Uruguay (1.66 mha), India (1.49 mha) and Germany (1.25 mha).

The world organic market is now 97 billion USD (F1BL & IFOAM, 2017). The growth of organic farming is approximately 20 per cent.

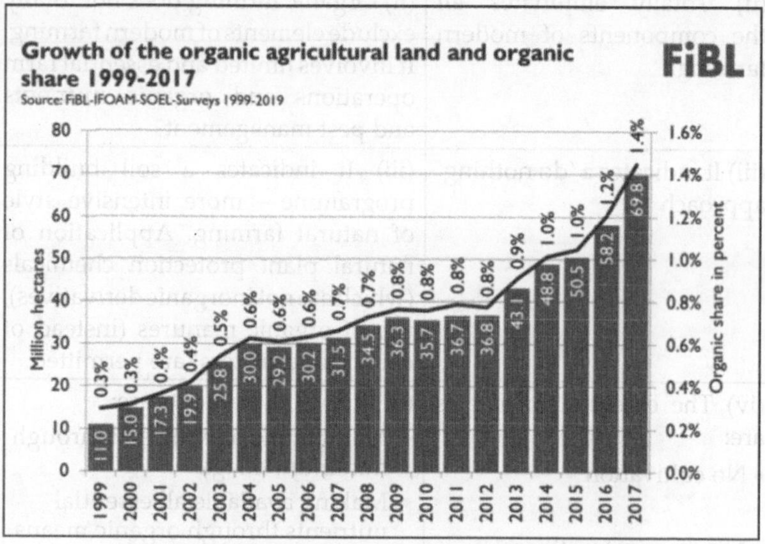

Fig. 1.1. World Growth of the Organic Farmland and Organic Share 1999-2017

Fig. 1.2. The Ten Countries with the Largest Areas of Organic Agricultural Land 2016

Area under certified organic cultivation in India was 42,000 ha during 2002-03 is increased to 3.43 mha during the year 2018-19, which includes certified cultivated area and wild certified area. The organic market from 1.15 million registered organic farmers in the country is valued at INR 51.51 billion with an annual growth rate of 12-15 per cent. India is becoming a major base for production and supply of organically produced agricultural products to the world market. India has developed National Standards under National Programme for Organic Farming (NPOP). Agricultural and Processed Food Products Export Development Authority (APEDA) and the National Centre of Organic Farming (NCOF) under Ministry of Agriculture is promoting organic entrepreneurs and farmers in India. In India leading states includes – Madhya Pradesh, Maharashtra, Rajasthan, Sikkim, Odisha, Gujarat, Uttar Pradesh and Uttaranchal. State wise farm area (excluding forest area) under organic certification is depicted in table 1.1.

Table 1.1. State wise Farm Area (including forest area) under Organic Certification Process in India (2018-19)

S.No.	State name	Organic area (in ha)	Rank
1	Andhra Pradesh	37,409	XII
2	Goa	20,964	XIII
3	Gujarat	94,708	VIII
4	Jammu & Kashmir	1,87,002	V
5	Karnataka	1,04,962	VII
6	Kerala	40,911	IX
7	Madhya Pradesh	9,18,303	I
8	Maharashtra	2,61,571	III
9	Odisha	1,27,851	VI
10	Rajasthan	6,32,701	II
11	Sikkim	75,798	IX
12	Uttar Pradesh	2,05,980	IV
13	Uttaranchal	41,409	X
14	Others	6,79,069	
	Total	**34,28,638**	

Source: Statistics for the year 2018-19, APEDA (https://apeda.gov.in)

1.7 COMPONENTS OF ORGANIC FARMING

The components of organic farming are as follows:

1.7.1 Organic Manures

Organic materials such as farmyard manure, biogas slurry, compost, straw or other crop residues, bio-fertilizers, green manures and cover crops can substitute for inorganic fertilisers to maintain the soil fertility and productivity. In addition, the organic farmers can also use sea weeds and fish manures and some permitted fertilizers like basic-slag and rock phosphate.

The use of organic manures will increase the organic matter content of the soil and thereby increasing its water holding capacity and reduce soil erosion. Crop rotation with legumes add to soil fertility. Green manure provides the nutrients and improves the soil.

1.7.2 Weed Control through Non-chemical Means

Compared to conventional farmers, the organic farmers use more of precautionary measures, good crop husbandary practices, mechanical and biological methods of weed control to reduce the weed menace. No herbicides are applied.

1.7.3 Biological Pest Management

The control of insect pests and pathogens is one of the most challenging jobs in tropical and sub-tropical agriculture. Here again non-chemical, biological pest management is encouraged. The conservation of natural enemies of pests is important for minimizing the use of chemical pesticides and for avoiding multiplication of insecticide resistant pests. Botanical pesticides such as those derived from neem based products could be used.

1.8 CONSTRAINTS

(i) Excessive cost of existing inspection and certification system which is not affordable by farmers.

(ii) Lack of quality assurance of organic inputs and non-availability of standards for their inputs.

(iii) Limited availability of locally available inputs like farmyard manure, compost, vermicompost, etc.

(iv) Limited domestic market and lack of commodity wise market information on domestic/global demand and supply.

(v) Non availability of organic package of practices for all crops based on locally available inputs.

(vi) Non-awareness of farmers and NGOs on the impact of organic farming on environmental pollution, health hazards, etc.

(vii) Risk of low production in initial years of organic farming adoptions due to slow release of nutrients from organic sources, not matching the nutritional demand of high-yielding varieties, and less effective pest and disease management options.

1.9 OTHER TYPES (FORMS) OF ORGANIC MANAGEMENT

1.9.1 Biodynamic Agriculture

Biodynamic agriculture is a method of farming that aims to treat the farm as a living system which interacts the environment, to build healthy, living soil and to produce food that nourishes and vitalizes and helps to develop mankind. The underlying principle of biodynamic is making life-giving compost out of dead material. The methods are derived from the teachings of Rudolf Stainer and subsequent practitioners. The important components of biodynamic farming are as follows:

- Turning in plant materials such as green crops and straw
- Not using chemical fertilizers and pesticides
- Avoiding soil compaction by machinery or animals, particularly in wet weather
- Keeping soil covered by pasture, crops or mulch not destroying the soil structure by poor farming practices such as excessive use of rotary hoe or cultivation in unsuitable weather (too wet or too dry)

- Fallowing the land by planting deep-rooting permanent pasture species or using green crops
- Use of preparations BD-500 and BD-501
- Compost made with preparations BD-502 – BD-507
- Liquid manure made with preparations BD-502 – BD-507
- Cowpat pit manure made with preparations BD-502 – BD-507

These biodynamic preparations named BD-500 to BD-507 are not food for the plants, but they facilitate the effective functioning of etheric forces. They are also not the usual compost starters, but can stimulate compost organisms in various ways. In short they are biologically active dynamic preparations which help in harvesting the potential of astral and etheral powers for the benefit of the soil and various biological cycles in the soil. So far 9 biodynamic preparations have been developed, named as formulation 500 to 508. Out of these, formulation-500 (cow horn compost) and formulation-501 (horn-silica) are very popular and are being used by large number of organic farmers. Formulations-502 to 507 are compost enrichers and promoters, while formulation 508 is of prophylactic in nature and helps in control of fungal diseases.

1.9.2 Rishi Krishi

Drawn from *Vedas*, the *Rishi krishi* method of natural farming has been mastered by farmers of Maharashtra and Madhya Pradesh. In this method, all on-farm sources of nutrients including composts, cattle dung manure, green leaf manure and crop biomass for mulching are exploited to their best potential with continuous soil enrichment through the use of *Rishi krishi* formulation known as "*amritpani*" and virgin soil. Fifteen kilogram of virgin rhizosperic soil collected from beneath of *banyan* tree (*Ficus bengalensis*) is spread over one acre and the soil is enriched with 200 litre *amritpani*. It is prepared by mixing 250 g *ghee* into 10 kg of cow dung followed by 500 g honey and diluted with 200 litre of water. This formulation is utilized for seed treatment (*beej sanskar*), enrichment of soil (*bhumi sanskar*) and foliar spray on plants (*padap sanskar*). For soil treatment it need to be applied through irrigation water as fertigation. The system has been demonstrated on a wide range of crops i.e. fruits, vegetables, cereals, pulses, oilseeds, sugarcane and cotton.

1.9.3 Panchgavya Krishi

Panchgavya is a special bio-enhancer prepared from five products obtained from cow dung, urine, milk, curd and *ghee*. Dr Natrajan, a medical practitioner and scientist from Tamil Nadu Agricultural University, Coimbtore has further refined the formulation suiting to the requirement of various horticultural and agricultural crops.

Panchgavya contains many useful micro-organisms such as fungi, bacteria, actinomycetes and various micronutrients. The formulation act as tonic to enrich the soil, induce plant vigour with quality production. Physico-chemical studies have revealed that *panchgavya* possess almost all macro and micronutrients and growth hormones (IAA, GA) required for plant growth. Predominance of fermentative micro-organisms such as yeasts and *Lactobacillus* helps to improve the soil biological activity and promote the growth of other micro-organisms. For foliar spray 3-4 per cent panchgavya solution is quite effective. Four to five sprays ensure optimum growth and productivity: (a) two sprays before flowering at 15 d interval, (b) two sprays during flowering and pod setting at 10 d interval and (c) one spray during fruit/pod maturation. Application of panchgavya has been found to be very effective in many horticultural crops such as mango, guava, acid lime, banana, spice turmeric, flower-jasmine, medicinal plants like coleus, ashwagandha, vegetables like cucumber, spinach, okra, radish and grain crops such as maize, greengram and sunflower. Panchgavya has also been found to be reducing nematode problem in terms of gall index and soil nematode population. As due to application of panchgavya a thin oily film is formed on the leaves and stem, it reduces evaporation losses and ensures better utilization of applied water.

1.9.4 Natural Farming

Natural farming emphasizes on efficient use of on-farm biological resources and enrichment of soil with the use of *jivamruta* to ensure high soil biological activity. Use of *Bijamruta* for seed/planting material treatment and *jivamruta* for soil treatment and foliar spray are important components. *Jivamruta* has been found to be rich in various beneficial micro-organisms. As per the studies conducted by Bio-centre, Bangalore, the *jivamruta* contains micro-organisms such as *Azospirillum* 2 x 10^6, PSM 2 x 10^6, *Pseudomonas* 2 x 10^2,*Trichoderma* 2 x 10^6,Yeasts and moulds 2 x 10^7 is applied at 500 litres *jivamruta* for one application in one hectare area. It can be applied through irrigation water by flow, by drip or sprinkler or even by drenching of mulches spread over the field or under the tree basin.

1.9.5 Natueco Farming

The Natueco farming system follows the principles of eco-system networking of nature. It is beyond the broader concepts of organic or natural farming in both philosophy and practice. It offers an alternative to the commercial and heavily chemical techniques of modern farming. Instead, the emphasis is on the simple harvest of sunlight through the critical application of scientific examination, experiments, and methods that are rooted in the neighbourhood resources. It depends on developing a thorough understanding of plant physiology, geometry of growth, fertility, and bio-chemistry. This can be simply achieved through: **`Demystification of Science'**.

Prayog Parivar has demonstrated that dissemination of relevant and often sophisticated science can be achieved in the local idioms of the common man. This can be very effective in bringing about a "gray matter revolution". With a new techniracy (technical literacy) for the management of soil, water, and canopy of leaves, it promises high yields with minimal external inputs and optimal harvesting of sunlight.

Natueco Farming methods go beyond natural farming and organic farming. In natural farming, farming is done trusting nature through the empirical wisdom of ages. However, Natueco methods emphasize farming by knowing nature more and more through critical scientific inquiries and experiments. It is an ever growing, novel, unique, participatory tryst between man and nature. Moreover, Natueco Farming in no way related to the present commercial techniques of farming. It has a new vision of infinite resource potentials in nature and sunlight and promises plenty for all through harvesting all available resources by increasing the human activity. This depends on critical understanding of greening and recycling of biomass within the neighbourhood to enrich the structure and fertility of soil in a calculated way. It promises record assured yields in a mathematic precision by understanding plant's geometry, cycles of growth and canopy (leaf area) management with little or no external inputs and ensuring optimum harvesting of sunlight. It visualizes that in the near future, the present money market system will have to give way to a new eco-economic system of nature, i.e. energy market system.

1.9.6 Homa Farming

Homa farming has its origin from Vedas and is based on the principle that "you heal the atmosphere and the healed atmosphere will heal you" The practitioners and propagators of *homa* farming call it a "revealed science". It is an entirely spiritual practice that dates from the *Vedic* period. The basic aspect of *homa* farming is the chanting of Sanskrit *mantras* (*Agnihotra puja*) at specific times in the day before a holy fire. The timing is extremely important. While there is no specific agricultural practice associated with *homa* farming, the farm and household it is practiced in, is energised and "awakened". The ash that results from the *puja* is used to energise composts, plants, animals, etc. *Homa* Organic Farming is holistic healing for agriculture and can be used in conjunction with any good organic farming system. It is obviously extremely inexpensive and simple to undertake but requires discipline and regularity.

Agnihotra is the basic *Homa* fire technique, based on the bio-rhythm of sunrise and sunset, and can be found in the ancient sciences of the *Vedas*. *Agnihotra* has been simplified and adapted to modern times, so anybody can perform it. During *Agnihotra,* dried cow dung, *ghee* (clarified butter) and brown rice are burned in an inverted, pyramid shaped copper vessel, along with which a special *mantra* (word-tone combination) is sung. It is widely believed that through burning organic substances in a pyramid-formed copper vessel, valuable purifying and harmonizing energies arise. These are directed into the atmosphere and are also contained in the remaining ash. This highly energized ash can successfully be used as organic fertilizer in organic farming.

Besides the practice of *Agnihotra* and the frugal distribution of *Homa* ash on beds and fields, a variety of further applications have also been recommended.

1.10 FUTURE PROSPECTS

India can emerge as global leader due to the presence of large number of organic producers (almost 5.85 lakh registered producers) and they needs to be supported with technical knowledge and inputs besides marketing infrastructure. The research available for little over a decade confirms the yield advantage in many crops such as basmati rice, maize, cotton, chickpea, soybean, groundnut, etc. However, the major impediment for growth of organic farming in India is yield reduction in the initial years due to swift switch over from inorganic to organic, wide gap between availability of organic source of nutrients requirement and lack of insect pests and diseases and weed management options. Most of the organic growers have expressed that lack of support price for organically grown crops and marketing infrastructure are the major constraints in promotion of organic agriculture. Although, much progress on research in organic farming has been done, the new emerging areas of human health benefits, understanding the economics with environmental markets, climate friendly farms and carbon farming with organic farming system models needs to be addressed in future. The certification systems of grower group, participatory guarantee system, know your farm and know your food should be promoted in large scale.

Chapter 2

Promotion of Organic Agriculture in India

Government of India has taken considerable efforts to promote the organic farming without compromising the food grain production. Organic farming produces a high premium food, normally priced 20-30 per cent higher than conventional food. In India more than 80 per cent of the farmers are small and marginal and many of them are practicing organic farming. However, since they are unaware of the market opportunity, they are not able to reap the benefits of organic farming. Hence, necessary steps need to be taken to address this issue.

Agricultural and Processed Food Products Export Development Authority (APEDA), New Delhi and National Centre for Organic Farming (NCOF), Ghaziabad (U.P.) promoting organic farming under Ministry of Agriculture in India. India is bestowed with lot of potential to produce all varieties of organic products due to its various agro - climatic regions. In several parts of the country, the inherited tradition of organic farming is an added advantage. This holds promise for the organic producers to tap the market which is growing steadily in the domestically and world over.

2.1 POTENTIAL AREAS OF ORGANIC FARMING

The organic farming perhaps is rarely practiced and there is dearth of data on relative yield and financial viability with organic farming verses intensive farming. Whatever little organic farming that is being practiced in India seems to be taken up mainly as part of contract farming and it may be promoted in selected rainfed and for export oriented crops. Soil and climatic conditions in India's rainfed areas make them practically well suited to organic agriculture. As far as potential areas are concerned three priority areas have been identified as follows:

- **Category I:** Areas where fertilizers and other agrochemicals consumption is very less. These areas are in Assam and other north-eastern states, Jharkhand, Orissa, Jammu & Kashmir, Himachal Pradesh, Karnataka, Madhya Pradesh, Chhattisgarh and Rajasthan.
- **Category II:** Areas under rainfed and hilly tract have tremendous opportunities for being converted into purely organic farming areas. Similarly, some areas under tea, coffee, spices and cashew may be easily brought under the organic farming with a trust on export of organic produce.
- **Category III:** Areas with irrigation and heavy use of fertilizers and other agrochemicals.

The appropriate strategy is needed to promote organic agriculture in the areas under category I and II on priority basis.

2.2 CHALLENGES

The Indian government has been undertaking measures to promote organic farming with the aim to improve soil fertility and help to raise the farmers' income. Some of the policy initiatives taken by government of India to promote organic farming and exports include development of an organic regulation for exports by the APEDA, removal of quantitative restrictions on organic food exports, providing subsidies to farmers under the *Paramparagat Krishi Vikas Yojana* (PKVY) in partnership with the state governments, and other schemes such as the Mission for Organic Value Chain Development for North Eastern Region. Despite these initiatives, following issues/ challenges to be addressed.

(i) **First, the supply chain is underdeveloped** and small and mid-sized farmers located in hilly regions and tribal belts find it extremely difficult to access the market. There is a shortage of pack houses and refrigerated vehicles, which leads to spoilage.

(ii) **Second, while the government is subsidising farmers** under the Participatory Guarantee System (PGS), which is a self-certification process supported through the PKVY scheme. These farmers are not allowed to export the produce. In fact, the APEDA has made it mandatory to have a third-party certification for exports.

(iii) **Third, as a farmer converts his/her land** from conventional chemical-based farming to organic farming, there is a risk of loss in yield due to the withdrawal of chemical inputs and high-yielding genetically modified varieties seed. However, in India, there is no such subsidy is given by the government during conversion period.

(iv) **Fourth, there is a serious shortage** of good quality organic inputs, which increases the risk of loss of yield. The available organic fertilisers are much below the required quantity, and there are a large number of spurious players in the market too. Similarly, there is a shortage of good quality organic seeds.

(v) **The fifth and the biggest challenge faced by organic** farmers is the lack of an organic policy for the domestic market and imports. In the absence of regulation on labelling standard for organic production and logo, it is not possible to distinguish an organic product from a conventional product. This has led to fraudulent practices and genuine players are not getting the premium, which the consumers of organic products are willing to pay.

2.3 STRATEGIES FOR PROMOTION OF ORGANIC FARMING

The strategy for promotion of organic farming includes identification of potential areas and crops. There is immense potential for promoting the organic farming in India particularly in rainfed areas. As regards crops, priority is for fruits, vegetables, spices, medicinal plants, oilseeds, pulses, cotton, wheat and basmati rice. Poor and marginal condition of soil as well as socio-economic condition of Indian farmers also support to promote it. For promoting organic farming in India there is need to give emphasis on the following areas:

(i) Development of Cropping System for Organic Farming: Development of suitable organic farming systems and practices in different agro-climatic and farming situations. These should be region specific particularly in rainfed areas, because in rainfed areas there is limited scope for use of chemical fertilizers. Development of package of practices of various cropping systems/intercropping/mixed cropping with high value crops based on organic inputs.

(ii) Technology for Promotion of Organic Farming: Recent development in technology like efficient use of crop residue, recycling of residue of dual purpose legumes, use of biomass of some of non conventional shrubs and trees, biogas slurry and vermi-compost for nutrient supply and use of bio-agent and predators for controlling pests and diseases to make comfortable promotion of organic farming in selected areas and high value crops.

(iii) Accreditation of Certified Agencies: Certification is a process of labelling in term that denote products have been produced in accordance with standards during food production, handling, processing and marketing stages and certified duly by a certification body. Certification and regulatory mechanism intended to assure quality and prevent fraud. For organic producers, certification identifies suppliers of products approved for use in certified operations. At present there are 31 accredited certifying agencies are operating in various places in the country.

(iv) Increase Awareness Among the Farmers: To increase the awareness among the farmers, conferences, seminars, and farmers' fairs may be organized. This will increase the knowledge of farmers about the organic on farm as well off farm inputs and related resources. Beside, appropriate research and extension system should be developed to make available, all the relevant information on organic farming in general and its specific technical details in respect to organic farming to the small and marginal farmers.

(v) **Government Subsidies and Provide Credit at Low Interest Rates:** Subsidy may be provided for organic inputs and produce in the form of development of production unit for composting, bio-pesticides, model organic farms and cold storage facilities at village level, etc.

(vi) **Promoting High Value Crops:** Lower yield of organic farming system can be compensated by the premium prices of organic produce. So that promoting the high value crops viz., horticultural crops, medicinal and aromatic plants, millets, basmati rice, etc. having export potential be given priority. Generally people are ready to pay premium prices for organic food. Hence, there is tremendous opportunity for promoting the high value crops.

2.4 SCHEMES TO PROMOTE ORGANIC FARMING IN INDIA

(i) **National Programme for Organic Production (NPOP):** The Government of India has implemented the National Programme for Organic Production (NPOP). This programme involves the accreditation programme for Certification Bodies, standards for organic production, promotion of organic farming, etc. The states promoting organic farming through this scheme are Kerala, Uttaranchal, Madhya Pradesh, Maharashtra, Karnataka, Gujarat, Rajasthan, Tamil Nadu, Sikkim, Nagaland and Mizoram. This programme provides information on organic production standards, criteria and procedures for accreditation of inspection and certification bodies. Standards and procedures have been involved in line with global standards like Codex. NPOP is administered under AGMARK by Ministry of Agriculture. NPOP standards for production and accreditation have been recognized by European Commission and Switzerland as well as USDA. With these recognitions, Indian organic products duly certified by the accredited certification bodies of India are accepted by the importing countries.

(ii) Paramparagat Krishi Vikas Yojana (PKVY): *"Paramparagat Krishi Vikas Yojana"* is an elaborated component of Soil Health Management (SHM) of major project "National Mission of Sustainable Agriculture (NMSA)". Under PKVY organic farming is promoted through adoption of organic village by cluster approach and PGS certification.

The scheme envisages: Promotion of commercial organic production through certified organic farming.

- The produce will be pesticide residue free and will contribute to improve the health of consumer.
- It will raise farmer's income and create potential market for traders.
- It will motivate the farmers for natural resource mobilization for input production.

This is a cluster based programme for promotion of organic farming. In this 50 or more farmers will form a cluster with 50 acre land to carry out organic farming under this scheme. In three years, around 10,000 clusters will form covering 5 lakh acre areas under organic farming; farmers will have no liability for expenditure on certification. As per the scheme, each farmer will be given INR 20,000 per acre in three years for seed to harvesting of crops and transport of produce to the market. Organic farming will also be promoted via traditional resources as part of the scheme. Linkages will be formed between organic products and the market as per this scheme. The scheme will also increase domestic production and certification of organic produce through the medium of farmers.

(iii) National Mission for Sustainable Agriculture (NMSA): This scheme works towards making organic farming sustainable, remunerative and climate resilient. Efforts are also on to develop natural resources and promote efficient water use "more crop per drop" for organic farmers. The scheme also works to develop the capacity of organic farmers and stakeholders in conjunction with other missions. It will also leverage other schemes such as MGNREGA, IWMP and RKVY. It will also establish departmental/ministerial coordination for deliverables under the scheme. It will work towards development of know-how and R&D for organic farmers.

The Mission specifically involves Soil Health Management (SHM) through residue management and organic farming practices. These are through: creation and linking of soil fertility maps with macro and micro nutrient management, relevant land use based on land use capability, judicious application of fertilizers and minimization of soil erosion/ degradation. The scheme will be implemented by State Government, National Centre of Organic Farming, Central Fertiliser Quality Control and Training Institute as well as Soil and Land Use Survey of India.

(iv) **Rashtriya Krishi Vikas Yojana (RKVY):** Assistance for promotion of organic farming on different components are also available under Rashtriya Krishi Vikas Yojana (RKVY) with the approval of State Level Sanctioning Committee. Under this following schemes are also funded:

- Mission for Integrated Development of Horticulture (MIDH),
- National Mission on Oilseeds & Oil Palm (NMOOP),
- Network Project on Organic Farming of ICAR.

(v) **National Project on Organic Farming (NPOF):** National Project on Organic Farming (NPOF) is a continuing central sector scheme since Tenth Five Year Plan. Planning Commission approved the scheme as PILOT project for the remaining two and half years of Tenth plan period with effect from 01.10.2004 with an outlay of INR 57.04 crore. The scheme is continuing in the Twelth Plan also.

NPOF is being implemented by National Centre of Organic Farming (NCOF) at Ghaziabad and its eight Regional Centres at Bangalore, Bhubaneshwar, Panchkula, Ghaziabad, Imphal, Jabalpur, Nagpur and Patna. Besides working for realisation of targets under NPOF, NCOF and RCOFs are also performing specific roles in promotion of organic farming.

The National Centre of Organic Farming, Ghaziabad, is conducting various extension and publicity activities like exhibitions, radio talks, TV programmes, distribution of literature on organic inputs / organic farming, etc. to educate farmers. NCOF is also publishing Bio-fertilizer News Letter as well as Organic Farming News Letter bi- annually and quarterly, respectively. In addition, various books on production and use of bio-fertilizers have been published in various languages for distribution to the farming community. Radio jingle (Hindi) on promotion of organic agriculture has been broadcasted on four different FM Radio Stations at Delhi, Banglore, Nagpur and Lucknow.

National Centre of Organic Farming is also organizing the training programmes for the farmers in *Saansad Adarsh Grams* (SAGs) from 2015-16.

Under National Project on Organic Farming (NPOF) scheme, assistance upto 25 and 33 per cent of financial outlay upto a ceiling of INR 40 lakhs and INR 60 lakhs is provided as back ended subsidy through NABARD for establishment of bio-pesticides/bio fertilizers production units and agro waste compost production units, respectively. Under NPOF, a total of 56 bio-fertilizers production units and 17 fruit/vegetables waste compost units have been established in the country. Government has been advocating integrating use of chemical fertilizers and organic manures including bio-fertilizers for increasing production of major crops.

(vi) **National Horticulture Mission (NHM):** Under National Horticulture Mission and Horticulture Mission for North East & Himalayan States (HMNEH), financial assistance is provided for setting up vermi-compost production units at 50 per cent of the cost subject to a maximum of INR 30,000/- per beneficiary for adoption of organic farming, INR 10,000/- per hectare for a maximum area of four hectares per beneficiary for organic farming certification and INR 5.00 lakhs for a group of farmers covering an area of 50 hectares.

(vii) **National Food Security Mission (NFSM)**: Under National Food Security Mission on Pulses, including Accelerated Pulses Production Programme (A3P), assistance for popularizing Rhizobium culture/phosphate solubilizing bacteria is provided to the farmers under cluster demonstrations.

(viii) **Schemes Promoting Bio-fertilizer Use:** Central Government has notified bio-fertilizers like *Rhizobium, Azotobacter, Azospirillum, Acetobacter*, PSB, KMB, zinc solubilizing bacteria under Fertilizer Control Order (FCO). Similarly, under Initiative for Nutritional Security through Intensive Millets Promotion (INSIMP) Programme, Phosphate solubilizing bacteria/*Azotobacter* culture is provided to the farmers as part of technology demonstration. Further, under National Project on Management of Soil Health and Fertility (NPMSH&F) financial assistance of INR 500 per hectare is provided to promote use of organic manures.

(ix) **ICAR Contribution in Promoting Organic Farming**: All India Network Project on Soil Biodiversity-Bio-fertilizers is implemented by Indian Council of Agricultural Research (ICAR) for R & D on bio-fertilizers. The ICAR has developed technologies to prepare various types of organic manures such as phospho-compost, vermi-compost, municipal solid waste compost, etc. Improved and efficient strains of bio-fertilizers specific to different crops and soil types are being developed under Network Project on bio-fertilizers.

The financial assistance is provided on the basis of project proposals received from States by Indian Council of Agricultural Research (ICAR) under Network Project on Organic Farming, with lead centre at Project Directorate for Farming Systems Research, Modipuram. This centre also developing package of practices of different crops and cropping systems under organic farming in different agro-ecological regions of the country.

In nutshell it can be said that for promotion of organic farming, identification of potential areas and crops is crucial. Strategies for promoting organic farming should be rainfed situation where agro-chemicals consumption is already very low and followed by rainfed farming with little scope of irrigation support. High value crops and local crops having export potential must be given priority. Technology like efficient use of crop residue, recycling of residue of dual purpose legume, use of biomass of some of non conventional shrubs and trees, biogas slurry, vermicompost, farmyard manure and leaf compost, etc. make organic farming comfortable promotion in selected areas and high value crops. Different schemes are operated by Govt. of India to promote organic farming in the country will definitely change the scenario of organic farming in the country in future.

Chapter 3

Organic Ecosystem and their Concepts

Organic agriculture faces the challenge and defiance in the present day pollution immersed environs. Soil, which is the source of about 90 per cent of our food, is the key component in any agro-ecosystem. Degradation creeps with the disability of survival of flora and fauna. Any constraint, if approached appropriately, can pave way to the formation and building up of agro- eco-systems, which can restore balance in the various cycles of water, nutrition, etc.

The monocausal approach of present day agricultural systems does not take account of the multifunctional nature and close inter linkages of the individual factors in the farm ecosystem. This has resulted in decreasing sustainability due to breaking of natural cycles of carbon and nutrient besides having several other deleterious effects on the environment which first affects the agriculture production but ultimately will become question mark for the human survival.

3.1 ORGANIC ECOSYSTEM

Agriculture is both a consumer and producer of ecosystem. A number of ecosystems are utilized to produce other ecosystems as food, which is supported by the maintenance of soil fertility, plant protection, water regulation and many other services. By using the concept of ecosystems, research and practitioners aspire to strike a balance between production and consumption of ecosystem in agriculture for long term farm sustainability.

Sustainable agriculture involves the use of natures' goods and services while maintaining them for future generations. Organic ecosystem is considered to be one of the production systems that aim to achieve sustainability by utilizing and maintaining eco-systems. The estimated magnitude of organic agriculture is very high compared to high input sustainable agriculture. It is well established that organic farming delivers more environmental benefits than does the conventional agriculture.

3.2 CONCEPTS OF ORGANIC ECOSYSTEMS

Eco-organic farming is not just substitution but rather adoption of eco-friendly and feasible technologies. Further, the revival and exploitation of local species of cultivable crops and cultivars, and also the lesser known or underutilized genetic resources is a vital link to build up sustainability.

Soil health care is fundamental to sustainable farming. In a healthy soil, the biotic and abiotic components covering organic matter including soil life, mineral particles, soil air and water exist in a stage of dynamic equilibrium and regulate the ecosystem processes in mutual harmony by complementing and supplementing each other when the soil is in good health, the population of soil fauna and flora multiples rapidly, which in turn, will sustain the bio-chemical process of dissolution and synthesis at a high rate. This state of soil life and the associated organic transformations will enhance the regenerative capacity of the soil and make it resilient to absorb the climatic vicissitudes and occasional failures in agronomic management. Since optimum plant nutrition is a key to any sustainable agriculture, the availability, transportation, treatment and application of biological materials must meet the major constraints in adoption of organic farming in present Indian conditions. Organic farming is a production system which favours maximum use of organic materials (crop residues, animal excreta, legume, on and off farm organic wastes, bio-pesticides, etc) and discourages the use of synthetically produced agro-inputs.

Alternate land use is defined as an effective economic utilization of land without harming the natural resources structures based on capability. The need of agro-forestry and other allied enterprises based organic farming is to ensure sustainability. A given system, particularly agro-ecosystem requires the support of forest component, which alone acts as the buffer. The restorative steps for reclamation of wastelands can be undertaken on a war footing to prevent the degradation in to wastelands and thus tree species perfectly suit to become a component in this systems management.

Ecological agriculture will reactivate the fatigued green revolution to move forward through promotion of agronomic practices with each component to a blend of tradition and modernity and a policy to value each input. It is sure that, all agricultural production of the future must come from the productivity of the soil, the base of agriculture. All farming practices have to be redesigned to undo the ill effects of intensive farming practices, for conservation and optimum utilization of all natural resources to ensure profitability with sustainability through indigenous technical knowledge and rationalize modern technology. Agronomic practices for dryland crops in India have been standardized, they now need to be dovetailed with efficient rainwater management and in-situ moisture conservation practices. The traditional conservation measures can be refined further to enhance the productivity.

3.3 IMPORTANT ASPECTS OF ORGANIC ECOSYSTEM

The most important aspects of organic ecosystem can be delineated as under:

(i) Focus on Multiple Commodities: It can enhance positive interactions between different biotic resources. This also prevents the monoculture bend and widens the scope of production, processing, marketing and myriads of other facets of the farming systems. Biological diversity has the potential for compensating among various components of the farming system. This brings in yield and economics, the current farming systems. This needs to be in terms of alternatives depending more on farm renewable resources. Fodder crop component particularly, summer legumes can find very significant position in many of the cropping systems and bring in resilience to the overall socio-economic standard of the peasants.

(ii) **Primarily on Farm Sufficiency, Orientation and Associated with Soil Fertility Build Up:** The technologies are developed for retuning nutrients from consumer to producer area. It is reported that there can gradual shift towards application of organic manures, particularly the crops residues and animal wastes, which are major source for composting manuring. Moreover, the vegetation on the bunds provides green manure based on availability. Thus the soil fertility was managed by lower external inputs and more from the on-farm produce itself.

(iii) **Regard for Environmental Balance:** Given the focus on the crop, field and at best farm level, the long term effects on soil fertility, the regenerative capacity of natural vegetation and fauna, human health, etc. require sufficient consideration. It is reported that biological methods were combined mechanical methods and plant communities with various layers and species were set up successfully and the "**red desert**" has been changed into a "**green ocean**". The concept of ecological agriculture in China had some difference as the practices involved use of limited external inputs which were safe and emphasis on the process of nutrient cycling, biological use of by - products and integrated pest management. The same is not restricted to small farms only, but can also be practiced at the level of a village, a township or a country even.

(iv) **Avoid Neglect of Rainfed Areas and Local Resources:** This aspect is particularly vital in marginal areas where Low External Input Agriculture (LEIA) farmers live and the local crops and animals provide their livelihood and rainfed agriculture need not continue to be gamble. The technologies for increased *in-situ* rain water retention, particularly, with increased vegetative cover even in cultivated lands, eg. the 800 to 1200 running meters of bunds including the boundaries can be covered with perennial fodders like *Styosanthes hamata* and other grasses locally available grass species help to stabilize bunds.

(v) **Due Value for Local Farmers Knowledge is Given:** Instead of the conventional top down approach, giving adequate importance to Indigenous Technical Knowledge

(ITK) to the farming community, ITK on different soil and water conservations. ITK is a people derived science and represents peoples creativity, innovations and skills. Further such techniques, which are changed over time through experimental and social changes.

(vi) Avoidance of Over Emphasis on Station- Based Research: Technology tested at station does work under farmer's condition, provided they are appropriately modified with an on-farm testing (OFT). There is a need to enmesh the ITK practices along with conventional soil and water conservation measures for promoting sustainable development of agriculture. The ITK may suitably modified/ refined for dissemination to new areas to increase the crop production.

3.4 ECO-ORGANIC FARMING DURING THE TRANSITIONAL PHASE

On studies regarding the transition experiences of some ecological farmers in South India, the farmers listed the following six essential points as under:

- The amount of organic matter in the soil must be increased in order to reduce dependence on synthetic fertilizers. This can be achieved by cultivating nitrogen fixing fodder crops and green manure and by increasing livestock numbers so that the supply of manure is increased.
- Soil tillage should be minimized and where possible replaced by mulching with cover crops, intercropping and the inclusion of trees in the fields.
- Weeds can be used as (living) mulch to prevent the evaporation of soil moisture and can also be used in preparing compost.
- A variety of selected trees should be planted to provide cattle fodder, improve the soil, supply green manure and save as windbreaks.
- Drought resistant species are preferable to annual crops and trees.
- Erosion control by contour bunding and soil cover is essential.

To conclude, despite the small troughs that may be faced, eco-organic has certain favorable effects, which can ultimately result in sustainability of the production system, includes improvement in soil heath (organic matter build up, favorable biological activities, favorable physical conditions, favorable chemical properties), biodiversity with emphasis on food production system, sustainability of food production system to a perceptible extent, improved landscape, etc. Organic farming is certainly a benign mode of achieving sustainability, provided communities actively participate in responsibilities, rationale and risks, to reap rights, receipts and recognitions.

Chapter 4

Organic Nutrient Resources and their Management

For growth and development, plant absorbs a large number of elements from the soil and other sources. At present twenty elements are considered essential for plant growth. The essential elements include carbon, oxygen, hydrogen, nitrogen, phosphorus, potassium, calcium, magnesium, sulphur chlorin, iron, manganese, boron, zinc, copper, molybdenum nickel, cobalt, sodium and silicon. These essential nutrients can be supplied to the plants by chemical fertilizers as well as through organic sources of nutrients of plant and animal origin. As inorganic farming use of chemical fertilizers is prohibited, it is implied that these essential nutrients required by the plants are to be supplied through organic means only. The organic sources of plant nutrients includes farm yard manure, compost, vermicompost, green manure, crop residue, concentrated organic manures and bio-fertilizers. To achieve sustainable production and productivity, maintain soil health, improve physical conditions of soil, it is necessary to use different nutrient sources in an integrated manner as per their availability and feasibility of application under different situations. The potential organic sources of plant nutrients are described as follows:

4.1 BULKY ORGANIC MANURES

They contain small percentage of nutrients and applied in large quantities. Farmyard manure, compost, vermicompost, crop residues and green manures are most important bulky organic manures. They have several advantages as (i) supply all essential plant nutrients (ii) improve physical conditions of soil (soil structure, water holding capacity, aeration, buffers soil surface temperature), (iii) increases availability of nutrients, (iv) stimulates soil flora and fauna, and (v) control parasitic nematodes and fungi to some extent by altering balance of micro-organisms in the soil.

4.1.1 Farmyard Manure (FYM)

Farmyard manure is the traditional organic manure. The term 'FYM' refers to decomposed mixture of dung and urine of farm animals along with litter and left over material from roughages or fodder fed to the cattle. FYM contains 0.5 % N, 0.25 % P_2O_5 and 0.5 % K_2O, on an average basis.

4.1.1.1 *Quality and Composition of FYM*

The quality and general composition of FYM is affected by the following factors:

(i) **Kind of Animal**: The quality of manure depends on the kind of the animal producing it. Sheep and poultry manures are richer in N, P_2O_5 and K_2O than cow, horse and pig manures. Within the same class, quality varies with individuality of animals. Some retaining more nutrients than others.

(ii) **Age and Condition of Individual Animal:** Growing animals, milch cattle, pregnant or carrying cattle utilize much of the ingredients in the feeds for building up their growing bodies, milk production and for the development of the embryo (calf). Old or adult animals kept on light work or no work, utilize little from feeds and as such, most of nitrogen is voided through urine and dung. Eventually, the adult old cattle provide better manure.

(iii) **Quality and Quantity of Feed Consumed:** Nutritious and protein rich feeds like oil cakes enriches the nitrogen content to the resulting manure than the bulky feeds like straw and green grass. Animals fed on concentrated feeds yield better quality manure.

(iv) **Kind of Litter Used:** The quality of manure depends to a considerable extent on the nature of litter used. Remnants of leguminous hays (*bhusa*) give richer manure than usual straws.

(v) Collection of Manure: The method adopted for collection of dung, urine and litter primarily decide the quality of manure as the loss of nutrients particularly nitrogen occurs from the time urine and dung are voided by cattle.The quality of manure depends upon the methods of collection *viz.*, Byre and Dry earth systems.

- **Byre system:** Cattle are stalled in a shed with a non absorbent floor provided with necessary slope towards the urine drains. The urine that flows into the drains is collected into a covered tank. From where it is periodically removed and sprinkled on the manure stored in a covered pit.The urine which is an important component of FYM can be properly stored (conserved) in this system. The perfect cleanliness and hygienic conditions of the stalls as well as cattle can be maintained in this system.
- **Dry earth system:** The floor of the cattle shed is well rammed and compacted. Layers of fine sand, red earth of loamy soil are spread as an absorbent for urine. The wet portions are properly covered with dry layers or any of the above materials and once a week the surface layer is removed and dumped in the manure pit. Available saw dust, paddy husk, groundnut shell, paddy winnowed dust would serve the purpose very well compared to the earth absorbents. This system is popular and extensively adopted in rural parts being cheap, convenient and practicable under the existing rural conditions in India.

(vi) Storage of Manure: Method of storage of manure influences the quality of manure to a large extent. During storage the manure undergoes fermentative changes, decomposition which leads to loosing its original structure and shape. There are three methods of storage viz., pit method, heap method and covered pit method. The details of different methods is given as under:

- **Heap method (above ground level):** This is the most common method adopted in Indian villages. Manure is heaped on the ground preferably under the shade of a tree. Ideal procedure is to dump the dung first and to cover it with litter soaked urine. This is further covered with a layer of litter/ash / earth to prevent the loss of moisture and to avoid direct exposure to sun. It is also desirable to put up a small bund around the base of the heap to protect against surface run-off washing out the manurial ingredients. It is beneficial to cover the exposed portion of the heap with palmyra leaves or any other available material. The maximum losses of nutrients occur in this method of storage, resulting in poor quality manure. Direct exposure to the vagaries of climate such as sunshine and rainfall causes looseness and dryness of manure, which hasten the losses of nutrients and rapid oxidation of organic matter.

- **Pit method (below ground level):** In this method, the manure is stored in a pit with non–absorbent bottom and sides. The pit is provided with a bund at the rim of the pit to prevent the surface run-off of waters during rainy season. The dimensions of the pit can be variable depending on the quantity of dung. The losses also occur in this method due to exposure to sun and rain, but it is relatively a better method than the heap method.
- **Covered pit method:** Of all the methods described, it is the best method. In this method, the bottom and sides of the pit are made non-absorbent by granite stone lining. The pit is also provided with a bund of 45 cm height to prevent surface flow of water (rain water) and a suitable cover by way of roofing with locally available materials like palmyra or phoenix leaves, etc. Organic matter and nutrient losses can be effectively controlled in this method of storage in order to obtain better quality manure.

4.1.1.2 *Improved Method for Preparation of FYM*

Trench method of preparing FYM advocated by C.N. Acharya is found one of the best method for preparing FYM. In this method the manure preparation should be carried in trenches of suitable size, say 6 to 7.5m length, 1.5 to 2.0m width and 1.0 m depth. All available dry litter and refuse from the farm and the houses should be heaped up near the cattle shed and portions of litter mixed with earth if available should be spread in the shed in the evening at 2.26 kg per animal for the absorption of urine. The litter should be localized in the areas where urine generally drops and soaks into the ground. Every morning the urine soaked litter and dung should be well mixed and then taken into the manure trench.A section of 90 cm length of the trench from one end should be taken up for filling with daily collection of refuse from cattle shed. When the trench is filled to the height of 45 to 60 cm above ground level, the top is made dome shaped and plastered with cow dung mixed with soil. The manure becomes ready by about three months. By this time the next 90 cm length of the trench being filled up. When the trench is completely filled up say in about three months, a second trench is taken for filling in the similar manner. By the time the second trench is filled up, the manure of first trench would be ready for use in 4-5 mo after plastering. Generally two such trenches would be sufficient to carryout the process round the year for a farmer having 3-4 heads of cattle. The chemical preservatives like gypsum, rock phosphate are added to reduce losses and enrich FYM. Bacteria and actinomycetes play active role in decomposition. Generally 60-70 per cent moisture in initial stage and 30-40 per cent moisture in decomposed manure (ready to use) as well as 50-60°C temperature under the heap are favourable for the activities of these micro-organisms. It is possible to prepare by this process 5 to 6 t of FYM per year per heads of cattle.

4.1.1.3 *Losses of Nutrients from FYM*

There are two types of losses of nutrients from FYM, which are as follows:

(i) Losses During Handling – FYM consists of two main components i.e. dung (solid portion) and urine (liquid portion) which are subjected to different losses in different ways. Both the components contain N, P_2O_5 and K_2O. Approximately, half of nitrogen and potash is in dung and other half in urine, whereas nearly all of the phosphorus (96%) is in the solid portion. To conserve N, P_2O_5 and K_2O, it is most essential that both the parts of cattle manure are properly handled.

- **Losses of dung:** Firstly, the most serious loss of dung is through dung cakes burnt as fuel for cooking. Besides, a large portion of cattle excrements is dropped outside the cattle shed, when the animals are grazing on the uncultivated lands. It is often said that about two third of total cow dung is either used for making cakes or lost during grazing, and only one third is used for making farmyard manure.
- **Losses of urine:** Urine contains nitrogen and potassium in large quantities. But there is no good method of preserving the urine in our country. In our country most of the cattle sheds have uncemented or *kachha* floor and the urine gets soaked in the soils of *kachha* floor of the cattle shed and the large quantities of nitrogen are thus lost as gaseous ammonia. The following reaction takes place:

(i) $$\underset{\text{(Urea in urine)}}{CO(NH)_2} + \underset{\text{(Water)}}{2H_2O} \rightarrow \underset{\text{(Ammonium carbonate)}}{(NH_4)_2CO_3}$$

(ii) $$(NH_4)_2CO_3 + 2H_2O \rightarrow \underset{\text{(Ammonium hydroxide)}}{2\,NH_4OH} + \underset{\text{(Carbonic acid)}}{H_2CO_3}$$

(iii) $$NH_4OH \rightarrow \underset{\text{(Ammonia) gas}}{NH_3\uparrow} + H_2O$$

The arrow above NH_3 indicates loss of ammonia as gas into the atmosphere.

(ii) Losses During Storage: Cow dung and other farm wastes so collected daily are accumulated in manure pit in open space for months together. The manure remain exposed to sun and rain and during this period nutrients are lost in the following ways:

- **By leaching:** Nutrients of farmyard manure are water soluble and these are liable to get lost by rain water. The leaching loss of nutrients will vary with the surface exposed to the intensity of rainfall and the slope of the surface on which manure is heaped. The leaching loss may be prevented by covering the pit by plastic sheet or roof over it.

- **By volatilization:** During storage, considerable amount of ammonia is produced in the manure heap from the decomposition of urea and other nitrogenous compounds of the urine, and from much slower decomposition of the nitrogenous organic compounds of the dung. During rotting process, a large quantity of ammonia is formed. This ammonia combines with carbonic acid to form ammonium carbonate and bicarbonate, which are rather unstable and gaseous ammonia may be readily liberated and passes into atmosphere.

(i)	$2NH_3$	+	H_2CO_3	$\rightarrow$	$(NH_4)_2CO_3$		
	(Ammonia)		(Carbonic acid)		(Ammonium carbonate)		
(ii)	$(NH_4)_2 CO_3$	+	$2H_2O$	$\rightarrow$	$2NH_4OH$	+	H_2CO_3
					(Ammonium hydroxide)		(Carbonic acid)
(iii)	NH_4OH	$\rightarrow$	$NH_3\uparrow$	+	H_2O		
			(Gaseous ammonia)				

The arrow above NH_3 indicates loss of ammonia as gas into the atmosphere.

4.1.1.4 *Ways to Minimize Losses of Nutrients*

(i) Adopt trench method of FYM preparation for handling of dung and urine.

(ii) Use of *Gobar* gas plant. The use of cow dung in bio-gas plant produces a combustible gas methane used as fuel gas which is an improved method of handling FYM.

(iii) Adopt covered method of storing FYM.

(iv) Adoption of BYRE system in collection of FYM.

(v) Spread the FYM in the field before ploughing, instead of piling of FYM in small heaps in the field.

(vi) Use of chemical preservatives like gypsum and rock phosphate at recommended doses i.e. 450 to 900 g per day per animal in the cattle shed where animal passes urine.

4.1.2 Compost

Compost *is a mass of rotted organic matter made from decomposition of plant and animal wastes with various additives*. The compost made from the farm wastes like weeds, paddy straw, sugarcane trash and other plants and wastes is called **farm compost**. The average nutrient content in farm compost is 0.5 % N, 0.15 % P_2O_5 and 0.5 % K_2O. The compost made from town refuses like street sweepings and dustbean refuse is called **town compost**. It contains 1.4 % N, 1.0% P_2O_5 and 1.4 % K_2O.

Composting is the natural process of 'rotting' or decomposition of organic matter such as crop residues, animal wastes, food garbage, etc by micro-organisms under controlled conditions.

The following basic rules are important for the production of good quality compost:

(i) The purpose of composting is to convert organic matter into growth promoting substances, for sustained soil improvement and crop production.

(ii) The organic matter is partially decomposed and converted by microbes. These microbes require proper growth conditions, for their activity i.e. moisture content: 50 per cent and 50 per cent aeration of total pore space of the composting material. This is achieved through stacking and occasional turning over. Microbes also need sufficient nitrogen for synthesizing their body cells, therefore the optimum C:N ratio of the composting material is 20:1 to 30:1.

(iii) Soil microorganisms constitute sufficiently to the decomposition of organic matter through their continuous activities. The majority of these soil animals provide optimal conditions in their digestive track for their synthesis of valuable permanent humus and stable soil crumbs. A typical compost earthworm is *Eisenia foetida*.

(iv) Certain additives accelerate the conversion and improve the final product. The materials such as lime, earth, gypsum, rock phosphate act as effective additives. The addition of nitrogen (0.1 to 1 %) is important in case of large C:N ratio of the composting material, provided it is not to be used in organic farming. The preparation of compost takes 2-3 mo. The composition of compost varies within wide limits.

4.1.2.1 *Methods of Composting*

There are several methods for preparation of compost, however, only some important ones are discussed here as under –

(i) **Activated Compost Method**: In this process the basic raw material i.e. straw and other farm wastes are treated with mixture of cattle dung and urine as decoction so that every portion of mass comes in contact with the inoculants (dung + urine) and fermentation takes place evenly. On piling up in a heap of about 1–1.5 m height and turning over from time to time, keeping moist with dung and urine decoction, very high temperature attained. When the temperatures begin to drop at the end of one week the volume of the material gets reduced. Further quantity of the basic material is added onto the heap. About 25 per cent of the new materials should be added at one time and thoroughly mixed with starters (dung +urine decoction) at intervals as before. If methodology is properly carried out, the compost will be ready in 5-6 wk.

(ii) **Indore Method:** The waste materials such as plant residues, animal wastes, vegetable wastes and weeds can be composted with the Indore Method. First the waste materials are chopped into small pieces of 5-10 cm size and are dried to 40-50 per cent moisture before stacking. Then they are spread in layers of 10-15 cm thickness either in pits or in heaps on 1 m width, 4-6 m length and 1 m depth. The heap is properly moistened with dung, using earth or night soil. Sufficient quantity of water is sprinkled over the heap to wet the composting materials to the level of 50 per cent moisture. Periodical turnings, usually three times at 15, 30 and 60 d after filling, are given to aerate and the material is covered with a thin layer of soil of about 2-3 cm thickness. Under the aerobic process of decomposition, losses of organic matter and nitrogen are heavy (40-50 per cent at initial stage). This process, however, involves considerable labour in the preparation of the heap and periodical turnings and so becomes expensive and impracticable when large quantities of materials are to be processed. The site selected must be at a high level to prevent rain water stagnation. It should be nearer to cattle shed and waste source for easy transporting.

(iii) **Bangalore Method (aerobic and anaerobic process):** This process of composting was developed by **Dr.C.N.Acharya in 1949.** The basic raw material is spread in a pit of 6 m length, 1.2 m width and 0.9 m depth to a depth of 15cm layer, moistened with 80–100 litre of water if the material is dry. Over this FYM or preferably a mixture of dung, urine and litter from the cattle shed is placed as a layer of 5 cm thickness. It is again covered on the top with a layer of earth to a thickness of 15 cm. It is beneficial to mix the earth with bone meal or oil cakes, wood ash, etc., to improve manurial value of the compost. The piling of layers is continued till the heap raises above the ground level to a height of 60 cm. Then the heap is kept open for one week to facilitate aerobic decomposition. Later the heap is plastered with a layer of moist clay for anaerobic fermentation to occur. The crakes that occur in the clay layer, have to be sealed off periodically. The compost will be ready in 4-5 mo period starting from the day of preparation. This process is called as aerobic and anaerobic decomposition of compost. In this process the basic raw material is not so well decomposed as in the other methods.But organic matter and nitrogen contents are well conserved. The number of turnings are reduced.The out turn of the compost is relatively greater and cheapest process.

(iv) **Coimbatore Method:** It is anaerobic degradation followed by aerobic process. First, pits of 4 m length, 2 m width and 1 m depth is formed in which crop residues or farm wastes are filled to a thickness of about 15 cm. Over this layer, cowdung slurry to enhance the rate of bio-degradation is applied to a thickness of 5 cm. Above this layer, 1 kg of bone meal or rock phosphate is applied to minimize the nitrogen loss and to add phosphorus. Thus, application of crop residues/ farm wastes, cowdung slurry, bone meal and rock phosphate in alternate layers is repeated till the height reaches 0.5 m above the ground level. Then the above ground portion is covered with red earth or mud to prevent the rain water entry and it becomes anaerobic process. After 30-35 d, the material is turned and it becomes an aerobic process. The compost will be ready within five months.

(v) **NADEP Compost:** Build a rectangular tank of 3 m length, 2 m width and 1 m depth made of brick walls and floor with mud plaster. Leave holes in the tank walls for aeration (about 4 holes along each side wall and two holes in each enclosure wall). Plaster the inner walls and the bottom of the tank with mud and cow dung mixture. For filling of compost pit put first layer of 15 cm of plant waste material followed by second layer of cow dung slurry (4 kg dung mixed with about 100 litres of water) and drench thoroughly. Then add 2 cm of sieved fine soil (60 kg) as third layer. Keep adding layers in this way until the materials are heaped 15 cm above the lip of the tank. Add another 7 cm layer of fine soil on the top heap and seal the tank with cow dung plaster (Fig. 4.1).

Fig. 4.1. NADEP Compost Pit

Build a temporary shed of thatch and bamboo to shield compost tank from direct sunlight and rain. After 3-4 mo., digestion is complete and compost is ready having dark colour and pleasant smell. Sieve through a thick mesh and use the compost. In order to improve the quality of compost, at 75-80 days make some holes by bamboo. Add 500 g PSB, 500 g *Azotobacter* and 500 g *Rhizobium* to 23 litres of water and pour through the holes and the holes are closed. This will increase the quality of compost. The disadvantage or limitation of this method is using large proportion of soil which is not desirable or acceptable in all situations.

4.1.3 Vermicompost

Among the soil fauna earthworms have attracted more attention than others because of their importance in agriculture. Earthworm's gut is the site of production of genuine humic acids which are distinct from the polysaccharide-gum humic acids. About half of the gums secreted by earthworms are in form of mucoproteins that help in stabilizing pore space distribution. The earthworm soil casts are richer in available plant nutrients (nitrate nitrogen, exchangeable Ca, Mg, K and P) and organic matter. The earthworms through their casts and dead tissues supply about 60-90 kg N to the soil. Earthworm eats on fungal mycelia. Earthworms convert farm wastes and organic residues into high quality compost. For this, *Eisenia foetida, Perionyx exacavatus, Eudrillus euginiae* and *Lumbrius rubellus* are important. These species can be cultured on organic wastes and dung. The technique of breeding and raising earthworms in controlled conditions scientifically is called **vermiculture,** and making compost with the use of earthworms is called **vermicomposting**. About 1,000 adult earthworms can convert 5 kg waste into compost per day. The earthworm assimilate 5-10 per cent of the substrate and rest passes through the alimentary canal and is excreted as cast. Earthworm cast contains nutrients, vitamins, hormones and antibiotics.

Vermicompost is a stable fine granular organic matter, when added to clay soil loosens the soil and provides the passage for the entry of air. The mucus associated with the cast being hygroscopic absorbs water and prevents water logging and improves water holding capacity. In the sandy soils where there is problem of water retention, the young strong mucus coated aggregates of vermicompost hold water for longer life. In the vermicompost, some of the secretions of worms and the associated microbes act as growth promoters along with other nutrients. It improves physical, chemical and biological properties of soil in the long run on repeated application. The organic carbon in vermicompost releases the nutrients slowly and steadily into the system and enables the plant to absorb these nutrients. The multifarious effects of vermicompost influence the growth and yield of crops.

4.1.3.1 *Types of Earthworms in Vermicomposting*

Earthworms belong to phyllum Annelida of Animal Kingdom. They are long and cylindrical in shape and size having a large number of grooves. There are about 3,000 species of earthworms in the world which are adapted to a range of environment. More than 300 species have been identified in India. Although hermophrodite, two mature earthworms are required to propagate. At the time of egg laying, the clitellum is transformed into hard, girdle like capsule called cocoon. Shedding of cocoon ranges from 1 to 5, only a few of them survive and hatch. The juveniles and again formation of cocoons takes a period of 50-60 d. Normally, the average life span of earthworms varies with species ranging from 1 to 10 yr.

Epigeics (surface feeders) are important in vermicomposting. The epigeics such as *Eisenia foetida* and *Eudrilus eugeniae* are exotic worms (Fig. 4.2) and *Perionyx excavatus* is a native one being used for vermicomposting in India. Epianecics are feeders on leaf litter and soil at upper layers of soil. The group such as *Lampito mauritii* is indigenous and is active in in-situ decomposition of organic wastes and residues in soil. Both epigeics and epianecics groups of earthworms are slender, shorter in length and red to dark brown in colour. They have high reproduction activity and efficient in recycling of organic materials. Increased attention has been paid to *Eisenia foetida* and *Eudrilus eugeniae* which have been found to be potential agent in vermicomposting of wide range of agricultural wastes and can grow at a wide range of temperature varying from 0-40º C. However, the optimum temperature ranges from 20-30ºC.

Eisenia foetida

Eudrilus eugeniae

Fig. 4.2. Epigeic (surface feeder) Exotic Worms

4.1.3.2 *Mechanism of Vermicomposting*

Materials consumed by worms undergo physical breakdown in the gizzard resulting in particles of size < 2 μ, giving thereby an enhanced surface area for microbial processing. This finely ground material is exposed to various enzymes such as protease, lipase, amylase, cellulase and chitinase secreted into lumen by the gut wall and associated microbes. These enzymes breakdown complex bio-molecules into simple compounds. Only 5-10 per cent of the ingested material is abosrbed into the tissues of worms for their growth and rest is excreted as cast. Mucus secretions of gut wall add to the structural stability of vermicompost.

4.1.3.3 *Methodology for Vermicompost Preparation*

It is mostly prepared in either pit or heap method. The dimensions either heap or pit are 3 × 1.2 × 0.6 m. The length and width can be increased or decreased depending on the availability of material but not the depth because the earthworms' activity is confined to 60 cm depth only (Fig. 4.3). First of all select a site which is not under any economic use and is shady (thatched roof or vermished) and there is no water stagnation. The site should also be nearby to water source. Put first layer of bedding material of 2.5 cm thick with soft leaves, then a second layer of 22.5 cm thick organic residue finely chaffed material and third layer of dung + water equal mixture of 5 cm layers. Continue the layers up to pile to ground level in the case of pit method and upto 50 cm in heap or surface bed method. Protect the worms against natural enemies like ants, lizards, snakes, frogs, toads, etc. by spray of neem based insecticides before filling the heap. Maintain proper moisture (60-70%), temperature (18-35°C) by turnings and subsequent staking. At the day of 24th, 4,000 worms are introduced into the pit ($1.0 m^2$=2,000 worms) without disturbing the pit. By regular watering the entire raw material will be turned into the vermicompost in the form of worm excreta. The turnover of the compost is 75 per cent of the total material accommodated in the pit, suppose 1,000 kg; the out turn will be 750 kg. On an average 1,000 earth worms may convert 5 kg material per day. The weight of 1,000 worms is approximately 1.0 kg.

Fig. 4.3. Vermicompost Pit

4.1.3.4 *Harvesting of Vermicompost*

Stop watering before one week of harvest. Sometimes the worms spread across the pit come in close and penetrate each other in the form of ball in 2 or 3 locations. Heap the compost by removing the balls and place them in a bucket. However, under most instances, top layer has to be disturbed manually. Earthworms move downward and compost is separated. After collection of compost from top layers, feed material is again replenished and composting process is rescheduled. The material is sieved in 2 mm sieve, the material passed through the sieve is called as vermicompost which is stored in a polythene bag.

Recomposting is done in the same pit or bed. Similar to the above described pit/heap method, vermicompost can be prepared in wooden box or brick column in similar way. In-situ vermicomposting can be done by direct field application of vermicompost at 5 t/ha followed by application of cowdung (2.5 cm thick layer) and then a layer of available farm waste about 15 cm thick. Watering should be done at an interval of 15 d.

Precautions

(i) Do not cover vermicompost beds/heaps with plastic sheets.

(ii) Do not overload the vermicompost heap to avoid high temperature that adversely affect their population.

(iii) Dry conditions kill the worms and water logging drive them away. Watering should be done daily in summer and every third day in rainy and winter seasons.

(iv) Addition of higher quantities of acid rich substances should be avoided.

(v) Make a drainage channel around the heap to avoid stagnation of water particularly in high rainfall areas in rainy season.

(vi) Organic materials used for composting should be free from non-degradable materials.

4.1.3.5 *Nutrient Composition*

Average nutrient composition of vermicompost is given in table 4.1 as under:

Table 4.1: Nutrient Composition of Vermicompost

S.No.	Nutrient	Content
1	Organic carbon	9.15 to 17.98 %
2	Total nitrogen	1.5 to 2.10 %
3	Total phosphorus	1.0 to 1.50 %
4	Total potassium	0.60 %
5	Ca and Mg	22 to 70 m. equi. / 100 g
6	Available S	128 to 548 ppm
7	Copper	100 ppm
8	Iron	1800 ppm
9	Zinc	50 ppm

Besides the above nutrients, vermicompost also contains protease, lipase, amylase, cellulase enzymes.

4.1.3.6 *Advantages of Vermicompost*

(i) Vermicompost is a rich source of nutrients, vitamins, enzymes, antibiotics and growth hormones, so it gives disease resistance to plants. Nutrient content of vermicompost is higher than traditional compost. It is a valuable soil amendment.

(ii) Vermicompost harbours certain microbial populations that help in nitrogen fixation and solubilization. Its application enhances nodulation in legumes and symbiotic mycorrhizal associations with the roots.

(iii) Superiority of vermicompost over other synthetic growth media is more pronounced in plant nurseries. It can be used as rooting medium and for establishment of saplings in nurseries.

(iv) It improves taste, lusture and keeping quality of the produce.

(v) It has immobilized enzymes like protease, lipase, amylase, cellulase and chitinase which keep on their function of bio-degradation of agricultural residues in the soil so that further microbial attack is speeded up.

(vi) It does not give foul odour as it is associated with manures and decaying organic wastes.

4.1.4 Vermiwash (a liquid manure)

Vermiwash a liquid collected after the passage of water through a column of worm action, very useful as a foliar spray to enhance the plant growth yield and infestation of diseases. It is a collection of excretory products and mucus secretions of earthworm along with nutrients from the soil organic molecules. Vermiwash, if collected properly is a clear and transparent, pale yellow coloured fluid. The utility of vermiwash in aquatic productivity was found to be immense but its potential as bio-cide either singly or when mixed with botanical pesticides is still under investigation. The vermiwash complex is observed to be efficient in raising of nurseries, lawns and orchards.

4.1.4.1 *Method of Preparation of Vermiwash*

A system of vermiwash preparation consist 30 cm height earthen or any other pot filled with pieces of stones upto 10 cm height from the bottom, above which a filter paper or muslin cloth or coarse sand is spread.

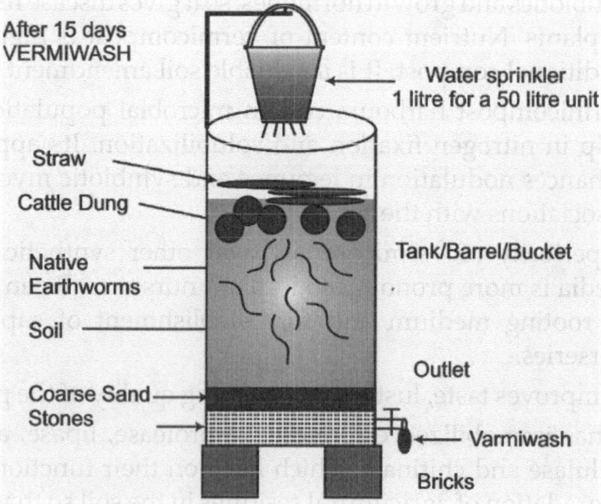

Fig. 4.4. Method of Vermiwash Preparation

Then a layer of soil and cattle dung along with 150-200 worms is laid down. The hole situated at the bottom of pot is fixed with a rubber hose, through which vermiwash is collected. Every day, two to three table spoons of fresh dung slurry was poured on the cattle dung as feed for the worms. Also about 100 ml of water is added everyday from the top. As it passes through earthworms burrows, a part of it washes down and the gut of the worms as also castings. Everyday 100 ml of vermiwash can be obtained by this method (Fig. 4.4).

4.1.5 Green Manuring

Green manuring can be defined as a practice of ploughing or turning into the soil undecomposed green plant tissues for improving physical condition of soil as well as soil fertility. From the time immemorial, the turning in a green crop for improvement of the soil conditions has been a popular farming practice. Green manuring, wherever feasible, is the principal supplementary means of adding organic matter to the soil. It consists of planting of a fast growing crop and ploughing it under to incorporate it into the soil. The green manure crop supplies organic matter as well as additional nitrogen, particularly if it is a legume crop, which has the ability to fix nitrogen from the air with the help of its root nodule bacteria. The green-manure crops also exercise a protective action against erosion and leaching.

4.1.5.1 *Advantages of Green Manuring*

(i) It adds organic matter in the soil, stimulates the activity of soil micro-organisms.

(ii) It improves structure of soil.

(iii) It decreases runoff, soil erosion and facilitates penetration of rainwater.

(iv) It holds plant nutrients, prevent leaching of nutrients.

(v) Leguminous green manure crops adds nitrogen to the soil (fixes 60 – 100 kg N/ ha).

(vi) It increases availability of plant nutrients like P, Ca, K, Mg & Fe.

4.1.5.2 *Disadvantages of Green Manuring*

(i) Under rainfed conditions, if adequate rains are not received after burying the green manure crop, it may affect proper decomposition and ultimately germination of subsequent crop and termite attack.

(ii) Loss of one crop, green manuring may not be always economical.

(iii) The cost of growing green manure crops may be more than the cost of commercial fertilizers.

(iv) An increase of diseases, insects and nematode is possible.

4.1.5.3 *Characteristics Desirable in an Ideal Green Manure Crop*

(i) Short duration, fast growing, high nutrient accumulation ability.

(ii) Legume with good nodular growth habit.

(iii) Less water requirement.

(iv) Can grow well on poor exhausted soils.

(v) Deep rooted, can open sub-soil and take nutrients from lower regions.

(vi) Leafy habitat capable of producing heavy tender growth early in its life cycle.

(vii) Large quantities of non-fibrous tissues for rapid decomposability containing fair percentage of moisture and nutrients.

4.1.5.4 *Types of Green Manuring*

Green manuring is done by two types, one is green manuring in-situ and another is green leaf manuring. The details are given as under–

(i) **Green Manuring In-situ:** In this system green manure crops are grown and buried in the same field which is to be green-manured, either as a pure crop or as an intercrop with the main crop. In India, various methods of growing green manure crops in-*situ* are followed to suit local conditions. For the proper decomposition of the green manure, it is necessary that the green material should be succulent and there should be adequate moisture in the soil. Plants at the flowering stage contain the greatest bulk of succulent organic matter with a low carbon/ nitrogen ratio. The incorporation of the green-manure crop into the soil at this stage allows a quick liberation of nitrogen in the available form. Green manuring in-situ is followed in Northern India. Important green manure crops are dhaincha, sesbania, sunnhemp, wild indigo, indigo and pillipesara. A brief about the cultivation practices of green manure crops is given as under:

- ***Sesbania aculeata* (dhaincha):** It is quick growing succulent green manure crop. It is suitable for loamy and clayey soils. Dhaincha is an ideal green-manure crop fits in different cropping systems. It is fairly resistant to drought as well as stagnation of water. The green manure grows well even in alkali soils and corrects alkalinity if grown repeatedly for 4-5 yr. The roots have plenty of nodules. The plant has soft stem. It makes good growth in 2-4 mo, and yields about 10 - 20 t of green manure per hectare. Stem gets woody and fibrous after three months of growth. As a pure crop it requires a seed rate of 25-30 kg per hectare.
- ***Sesbania speciosa* (sithagathi):** It is a valuable green manure for wetlands and can be grown in a wide range of soils. There are different methods of growing *Sesbania speciosa* in rice field. Three to five days prior to harvest of rice crop, seeds at 50 kg per hactare are sown as broadcast. These seeds get mixed into the soil while trampling of labours during harvest. With the available soil moisture the seeds of *Sesbania* germinate. In other methods, after ploughing the field, *Sesbania* seeds are broadcasted at 35-50 kg per hactare. A good stand of crop can be obtained by irrigation. Where two crops of rice are taken, the green manure can be raised on the field borders along the bunds. Sesbania seedlings (21d) can be planted in a single line at 5-10 cm apart in the borders of the fields close to the bunds. In about 90 d it produces about 2-4 t of green manure per hectare. It does not affect the rice yield by shading or root

effect. If second rice crop is planted immediately after the first crop, the manure can be incorporated into the field. About 300-400 g of seeds are sufficient to raise nursery and plant the seedlings along the bunds of one hectare.

- ***Sesbania rostrata* (sesbania):** It is an aquatic leguminous crop which has nodules both on stem and roots. The stem nodulation is an adaptation for waterlogged situation since flooding limits growth of green manures and may reduce root nodulation. Under normal conditions both root and stem nodules are effective in nitrogen fixation. For growing pure crop the normal seed rate is 30 to 40 kg per hactare. To get early, uniform germination and vigorous seedling, seed have to be scarified with concentrated sulphuric acid for 15 min. It can produce about 40-50 t of fresh biomass from one hectare in about 50-60 d. It has higher nitrogen content of 3.56 per cent on dry weight basis. Bio-mass production is higher during summer (April-June) than in winter (December-January) season. This green manure can also be raised by raising seedlings (30 d old) and planted in the paddy field along the bunds or as intercrop with rice.
- ***Crotalaria juncea* (sunnhemp):** It is a quick growing green manure crop and gets ready for incorporation in about 45 d after sowing. It comes well in loamy and heavy soils. It does not withstand heavy irrigation or continuous water logging. The seed rate is 25-40 kg per hactare. The crop can produce about 10-20 t of green bio-mass per hactare depending upon the environmental conditions and duration of crop.
- ***Tephrosia purpurea* (wild indigo):** It is a slow growing green manure crop and cattle do not prefer to graze it. The green manure is suitable for light textured soils, particularly in single crop wetlands. If the crop is sown for two to four seasons in the same field it becomes self sown in subsequent years and thereafter there is no need for fresh sowing of seeds in the same field. On an average about 3-4 t of green manure is obtained from one hectare area.

The seed rate is 30-40 kg per hactare. The seeds have a waxy impermeable seed coat and hence scarification required to induce germination. Soaking of seeds in boiling water for 2-3 min is also equally effective in promoting germination.

- ***Indigofera tinctoria* (indigo):** It is a perennial shrub. It is found wild and in cultivated lands. There are two types which closely resemble each other are Madras indigo and Bengal indigo. It comes up well in clayey soils with one or two irrigations. The seed rate is 25 to 30 kg per hactare and yield of green matter varies from 10-12 t per hactare.
- ***Phaseolus trilobus* (pillipesara):** This is a dual purpose crop yielding good fodder for the cattle and green manure. Pillipesara comes up well in hot season with sufficient soil moisture. Loamy or clayey soils are best suited. Initially, adequate soil moisture is essential for its early growth. One or two irrigations given during its growth period will help in producing bumper harvest of forage crop. After taking one or two cuttings for fodder or light grazing by animals, the crop can be incorporated into the soil. The seed rate is 20-25 kg per hactare. About 8-10 t of manure can be obtained from one hectare area.

(ii) Green Leaf Manuring: Green leaf manuring refers to turning into the soil green leaves and tender green twigs collected from shrubs and trees grown on bunds, waste lands and nearby forest areas. A number of leguminous and non-leguminous plants are grown on bunds or wastelands with the prime objective of utilizing their foliage as green manure. It is mostly practiced in eastern and central India. The common shrubs and trees used for green leaf manuring are described as under:

- ***Glyricidia maculata* (glyricidia):** This is a shrub type plant that comes up well in moist situations. Under favourable conditions, it grows well like a tree. It can be easily grown in waste lands, farm road sides, field bunds, etc. The crop can be established by stem cuttings or seedlings planted in the field borders. It can be pruned for its tender loppings and compound leaves for green leaf. Established each plant yields 12-15 kg green leaf. About 400 plants on the peripheral bunds yields 5-6 t green leaf per hectare. Within two years after planting the plants are ready for lopping. The shrub can be pruned two to three times a year and it withstands repeated loppings. When the shrubs are regularly lopped the height to be restricted to 2-3 m and they do not affect the growth of cultivated crops with their shade effect.

- ***Pongamia glabra* (karanj):** It is a leguminous tree grows on coastal forest, on river banks and on tank bunds mostly along streams, wastelands and road sides. The trees are established by means of planting two to three months old seedling, 4-5 m apart. Lopping may be taken once or twice a year on trees grown in wastelands. On an average, a tree can yield 100-120 kg of green matter. The leaves contain about 3.7 per cent nitrogen (on dry weight basis).
- ***Azadirachta indica* (neem):** It is an indigenous tree with wide range of adaptability. It has profusely branching habit and green tree gives plenty of foliage. It can be raised in all types of soil. The trees are grown on field boundaries, river banks, roads, wastelands at 5-6 m distance. The leaves are pruned at the time of planting rice and incorporated as green leaf manure. One or two loppings in a year are taken in favourable seasons. Each lopping gives about 150 to 200 kg of leaf matter.
- ***Ipomoea carnea* (Besharam):** It is a quick growing, profusely branching, and highly drought resistant weed. It gives abundant green leafy material in short time. It is multiplied by means of mature stem cuttings. Stem cuttings of about 0.30 m long with three or four nodes and axillary buds are planted at a distance of 1 to 2 m all along the wide field bunds, irrigation channels and fences. As many as 1,800-2,000 cuttings can be accommodated in one hactare area as border planting and two to three loppings can be taken in a year. Each plant will give about 5 to 7 kg of green matter per lopping.

- ***Cassia auriculata* (Avaram senna):** It is a very common plant, found coming up in waste lands, hill slopes, plain sea shores, etc., almost in the wild condition. It is a hardy plant. The plant is propagated through seeds. The seeds get dispersed and plants grow naturally without any efforts. When the plants start to flower in off-season, they are cut and applied in the fields.
- ***Thespesia populnea* (Portia tree):** It is also an ever-green tree which thrives in all types of soils. The trees are grown in garden land areas, gardens and also in waste lands. A spacing of 4 to 5 m is adopted. It is propagated by stem cuttings. It establishes very quickly and produces a number of branches. Two or three loppings of green leaves are taken in a year during favourable seasons. A tree will give as much as 100 to 150 kg of green matter per lopping.

4.1.5.5 *Stage of Incorporation*

In in-situ green manuring the best stage of incorporation is flowering stage of the crop. In green leaf manuring incorporate the leaf before mature or attain woody nature. Plants at very young stage may not be incorporated as they will decompose very easily, leaving little residue in the soil.

4.1.5.6 *Time of Incorporation*

It is applied as basal dressing before main crop is raised in the field. Usually about 6 – 8 wk time is found to be sufficient for decomposition. In fruit crops, plantation crops, sugarcane, etc. green manure crop is grown along the planting of main crop and incorporated after 40-50 d of growth at the time of earthing or intercultural operation. The plants /green leaf are trampled by working the implement and later on levelling the land. In some cases when it is grown along the crop it is incorporated with intercultivation like in sugarcane.

4.1.6 Crop Residues

The contribution from crop residues is generally ignored. However, crop residues add considerable amount of nutrients and the amount depends on the crop. For example finger millet crop residues add about 43 kg N/ha, while rice crop residues add 17 kg N/ha. The addition of phosphorus is 3.7 and 2.9 kg P_2O_5/ha by finger millet and rice residues, respectively.

Large amounts of agricultural wastes are available and there is need to properly use them for some industrial purpose or recycle them to replenish soil fertility. Agricultural residues like paddy straw, rice husk, jute/cotton/pigeonpea sticks, wheat straw, groundnut shells, maize stover, sugarcane trash, etc. are left in the fields which create many ecological problems. Perpetuation of diseases and other insect pests is helped by such wastes/residues. For example, cotton sticks left in the field harbour pests to affect the next season cotton crop seriously. Burning of rice husk near rice shellers, in the vicinity of towns, creates air pollution. Burning of paddy straw by farmers in their fields also gives out lot of smoke adding in pollution leading to enhanced incidence of deaths due to increased respiratory diseases.

Industrial wastes like rice husk, rice bran, bagasse, press mud, cotton dust, oil cakes, slaughter house wastes, faecal matter, marine residues, city garbage, night soil, etc. could also provide NPK equivalent to about 4.72 mt for increasing agricultural production and, thus, become a vast source of energy after developing a sound system of their use and recycling.

A large quantity of about 247 mt of agricultural by-products like wood, dung and crop residues are burnt annually in India due to fire needs. This vast source of energy, thus, could not be recycled back into the ecosystem, until an alternative source of energy is provided for home consumption. Biogas technology may help to provide valuable energy source for cooking and would also provide digested slurry as a valuable source of organic manure. Systematic, efficient and safe collection and disposal of city garbage like rubbish, dead animals, tanneries wastes, street sweeping, etc. could be economical, besides, reducing environmental pollution, foul smell and nuisance for health hazard. With proper treatment such garbage could be turned to valuable compost.

4.1.6.1 *Crop Residue Generation*

After crop harvesting, the left over plant material including leaves, stalks and roots is known as crop residue. It is estimated that India generates approximately 500 mt of crop residue annually as per Govt. of India, 2016 statistics. Cereals, fibers, oilseeds, pulses and sugarcane contributed the majority of crop residue with production estimates of 352, 66, 29, 13 and 12, mt, respectively (Table 4.2). Among cereal, crops, rice, wheat, maize, and millets together contributed (70%) of crop residue followed by fiber crops (13%).

Table 4.2. Yearly Average Crop Residue Generation in India (2010-14).

Crop	Crop reside generations deviation (million ton)
Rice	121.2 ± 17.04
wheat	114.14 ± 11.56
Coarse cereal	71.77 ± 7.09
Pulse	27.99 ± 3.14
Crop	**Crop reside generations deviation (million ton)**
Oilseed	33.04 ± 4.3
Sugarcane	137.88 ± 8.18
Cotton	11.4 ± 1.05
Other fibres (Jute & Mesta)	3.8 ± 0.47
Total yearly average	**521.1** ± **52.83**

4.1.6.2 *Environmental Pollution*

Typically, surplus residue is burnt in- situ from March to May. On farm burning of crop residue, becomes a source of green house gases (Co_2, Co, CH_4, N_2O, SO_2, aerosols, particulate matters, smoke, volatile organic compounds and radioactive gases, major source of environmental pollution. In view of this it is very important to use this precious organic nutrients resource for crop production especially in organic farming.

4.1.6.3 *Utilization of Crop Residue*

Imposing a ban on burning of crop residue may not be fruitful unless growers are enlightened with the negative effects of it on human, animal and soil health, crop biodiversity, the micro- and macro-environment, etc. To disseminate the knowledge of usefulness of crop residue, extension education is encouraged among growers and producers. Extension activities include talks, speeches, visuals, presentations, publication, etc. For example, a documentary on the environment and climate change with special emphasis on how burning crop residue adversely impact on climate change can be used. Educating the farmers about the pecuniary and non pecuniary benefits of not burning the agricultural residues could include, alternatives to burning agricultural residue like collection and transportation of agricultural residues, gasification as a fuel for the boilers, converting into briquettes, designing of suitable harvester, composting *in-situ,* and straw mulching while using disc ploughs, disc harrows, rotavators, zero tillage, etc.

4.2 CONCENTRATED ORGANIC MANURES

Concentrated organic manures are those materials that are organic in nature and contain higher percentage of essential plant nutrients such as nitrogen, phosphorus and potash, as compared to bulky organic manures. These concentrated manures are made from raw materials of animals or plant origin. The concentrated organic manures commonly used are aquatic weeds, oilcakes, blood meal, fish meal, meat meal and horn & hoof meal.

4.2.1 Oil Cakes

A large variety of oil cakes are produced in the country. Oil cakes can be grouped into two categories, namely, (i) Edible oil cakes suitable for feeding the cattle and (ii) non-edible oil cakes – not suitable for feeding the cattle. Edible oil cakes are normally fed to cattle as concentrates. However, some of the edible oil cakes are also applied by the farmers to the soil. The quantity of nitrogen varies with the type of oil cake. It is from 2.5 per cent in mahua cake to 7.9 per cent in decorticated safflower cake. In addition to nitrogen, all oil cakes contain small percentages of P_2O_5 (0.8 to 2.9 %) and potash (1.2 to 2.2 %). Average nutrient content of different non edible and edible oil cakes is given in table 4.3. Oil cakes are the quick acting organic source becomes quickly available to the plants in about 7 to 10 d after application. The mahua cake, however is an exception, as its nitrogen does not become available till about two months after application. As such mahua cake should be applied about two months before sowing, provided the soil is moist. This cake is suitable for application to fruit plants and plantation crops.

Table 4.3. Average Nutrient Contents of Different Oil Cakes

Oil cake	Nutrient content (%)		
Non- edible	**N**	**P_2O_5**	**K_2O**
Castor cake	4.3	1.8	1.3
Cotton seed cake. (undecorticated)	3.9	1.8	1.6
Karanj cake	3.9	0.9	1.2
Mahua cake	2.5	0.8	1.2
Safflower cake (Undecorticated)	2.9	1.4	1.2
Oil cake	**Nutrient content (%)**		
Edible			
Coconut cake	3.0	1.9	1.8
Cotton seed cake (decorticated)	6.4	2.9	2.2
Groundnut cake	7.3	1.5	1.3
Linseed cake	4.9	1.4	1.3
Niger cake	4.7	1.8	1.3
Rape seed cake	5.2	1.8	1.2
Safflower cake (decorticated)	7.9	2.2	1.9
Sesamum cake	6.2	2.0	1.2

For field application, oil cakes should be well powdered so that they can be spread evenly and are easily decomposed by micro-organisms. They can be applied a few days prior to sowing or at sowing. Oil cakes especially groundnut cake, are also applied extensively in the form of a top dressing to sugarcane. Depending on crop, oil cakes are applied broadcast, drilled or placed while earthing up near root zone.

4.2.2 Meat - meal

An adult animal can provide 35 to 45 kg of meat-meal after slaughter or death. At present, meat-meal is manufactured in India on a small scale. The process of manufacturing meat- meal is very simple. First the bones and the meat are cooked or digested in a special container for two to three hours. The bones are then separated from the meat. This meat is dried on a sand bath till it is brittle and then it is powdered. The drying can be done in double-jacket trays worked by steam and the material dried over the steam for an hour. Mutton squeezer can also be used for removing water content of the meat, and then dried. Meat-meal is a quick acting manure and is effective for all crops, on all soil types. Its application is similar to oil cakes. Meat-meal contains about 10.5 per cent nitrogen and 2.5 per cent phosphate.

4.2.3 Blood - meal

A adult cattle gives about 14.0 kg, and a goat or sheep about 1.40 kg of blood. Slaughter houses should be provided with pucca or concrete floor with a central drain leading into a blood storage tank. The blood is first treated with commercial copper sulphate at 125 g per 100 kg of blood. It is then evaporated to dryness on a sand bath. Next, it is spread on a concrete floor covered by a net, and allowed to dry in sun. When completely dried, it is powdered, bagged and marketed as blood-meal. Blood-meal is quick acting manure and is effective for all crops on all soil types. It should be applied like oil cakes. Blood-meal contains 10 to 12 per cent of nitrogen and 1 to 2 per cent phosphate.

4.2.4 Fish - meal

Non – edible fish carcasses are used to prepare fish-meal. The fish is dried, crushed or powdered and filled in bags. Fish manure is available either as dried fish or as fishmeal or powder. The manurial constituents present in it vary with the type of fish meals. It should preferably be powdered before use. Fish-meal contains 4 to 10 per cent nitrogen, 3 to 9 per cent phosphate and 0.3 to 1.5 per cent potash.

4.2.5 Horn and Hoof meal

A stout animal gives about 3-4 kilogram of horn and hoof. These materials, after cooking in the bone digester, are taken out, dried and powdered. Such a meal contains about 13 per cent nitrogen.

4.3 BIOFERTILIZERS

Biofertilizers or microbial fertilizers or more appropriately microbial inoculants' are preparations containing live or latent cells of efficient strain of nitrogen fixing micro-organisms used for seed or soil application with the objective of increasing the numbers of such micro-organisms in soil or rhizosphere and consequently improve the extent of micro-biologically fixed nitrogen for plant growth. They are used either to fix nitrogen or to solubilize plant nutrients like phosphates or to otherwise stimulate plant growth through synthesis of growth promoting substances or to collect available phosphorus from remote places out of reach of plant root hairs by sending elongated filaments (VAM).

It is of utmost importance to incorporate biofertilizers in integrated nutrient management strategy for different crops and cropping system in organic farming as it very useful and cost effective technology available to crop production as nutrient supplement.

4.3.1 Classification of Bio-inoculants/Biofertilizers

The bio-inoculants can be classified as symbiotic nitrogen fixing organisms, non - symbiotic or free living organisms and other bio-inoculants. The details of these are given as under.

4.3.1.1 *Symbiotic Nitrogen Fixation*

Bacteria belonging to genus *Rhizobium* are capable of fixing atmospheric nitrogen in association with legume crops. *Rhizobium* spp. enter the roots of host plants and form nodule on the root surface. The bacteria depend on the host plant for carbohydrates and water. *Rhizobium* supplies nitrogen to the host. Nitrogen fixed by the *Rhizobium* is translocated through xylem vessels of the host plant mainly in the form of aspergine and to some extent as glutamine. The colour of nodule varies from whitish to brown, green and pink. The pink coloured nodules due to the presence of pigment legheamoglobin are called effective nodules. This pigment helps in nitrogen activity in process of N fixation. Rhizobium species suitable for different legume crops is given in table 4.4

Table 4.4. Rhizobium Species Suitable for Different Crops.

Rhizobium spp.	**Crops**
R. Leguminosarum	Peas (*Pisum*), lathyrus
R. Trifoli	Berseem (*Trifolium*) and clover group
R. Phaseoli	Bean group (*Phaseolus*)
R. Lupini	Lupine group
R. Japonicum	Soybean (*Glycine*)
R. Meliloti	*Melilotus* spp. lucerne (*Medicago*), fenugreek (*Trigonella*)
Cowpea miscellany	Cowpea, clusterbean, greengram, blackgram, redgram, groundnut, kidneybean, dhaincha, sunnhemp, *Glyricidia, Acacia, Prosopis, Dalbergia, Albezzia, Indigofera, Tephrosia, Atylosia, Stylo*, etc.

The quantity of nitrogen fixed by different legumes is depicted in table 4.5 as under:

Table 4.5. Quantity of N Fixed by Important Legumes (range).

S.No.	**Crop**	**Nitrogen fixed (kg/ha)**
1	Alfalfa	100-300
2	Blackgram / greengram	50-55
3	Clover	100-150
4	Clusterbean	37-196
5	Chickpea	85-100
6	Cowpea	80-85
7	Fenugreek	45-50
8	Groundnut	50-60
9	Lentil	90-100
10	Pea	52-57
11	Pigeonpea	168-200
12	Soybean	60-88

Method of Application

Biofertilizers can be applied to different crops and plants by three different ways.

- **Seed Treatment:** Carrier based inoculants are mixed with little water for preparing slurry in which small quantity of sugar or jaggery or gum is added so that the inoculants may get energy for their prolonged survival. The slurry of inoculants should be thoroughly mixed with the seeds. Now the treated seed should be dried in cool and dry shady place and sown immediately in the field. In case of acidic soils, pelleted seed with inoculants should be sown. Pelleting agents used are lime, rock phosphate, charchol, gypsum, etc.
- **Soil Treatment:** The *Rhizobium* inoculum is mixed with soil and spread over field.
- **Soil Application:** If *Rhizobium* inoculum is not available 200 kg of surface soil (2-10 cm) is taken from the field where that particular crop was sown in last season, and broadcast in the field where the crop is to be sown.

4.3.1.2 *Non Symbiotic or Free Living Organisms*

(i) **Azotobacter:** It is heterotrophic free living nitrogen fixing bacteria present in neutral and alkaline soils. *Azotobacter chroccum* is the most occurring species in arable soils. It is also known to synthesize biologically active growth substances such as Indole Acetic Acid (IAA), Gibberellins, B-vitamins, nicotinic acid, pantothenic acids, biotin and heteroauxin. Application of 5 t per hectare FYM helps in better growth of *Azotobacter*. It is applied by seed inoculation or seedling dip or soil application. The inoculum required at 3-5 kg per hectare. It fixes 20 – 30 kg per hectare nitrogen. It can be applied in maize, wheat, rice, cotton, sugarcane, etc.

(ii) **Azospirillum:** *Azospirillum lipoferum* and A. *brasiliense* are the important species. This organism forms associative symbiosis with many plants particularly with those having C_4 pathway. It is recommended for sorghum, corn, *Panicum, Setaria,* etc. A reduction of 25 per cent nitrogen requirement of crops similar to that of *Rhizobium*. It is applied through seed treatment at 600 g per hectare. For field application 2 kg culture + 25 kg FYM + 25 kg soils per hectare is required. Increase in yield is mainly due to release of growth promoting substances and protection from diseases. Optimum temperature is 32-35° C for good growth of *Azospirillum*.

(iii) **Blue Green Algae (BGA):** It is a photosynthetic nitrogen fixing organism. This algae is congenial to rice ecosystem and it is often referred to as **"Paddy organism"** because of its abundance in paddy fields. It uses sunlight as source of energy and water as a source of reductant for photosynthesis and nitrogen fixation. Nitrogen fixing BGA mainly belongs to orders *Nostocales* and *Stigonemetales.* Blue green algae may be in higher plants such as tiny aquatic fern azolla where *Anabena azolla* propagates in symbiotic association. It is a potential source of organic manure and nitrogen in rice production. BGA also liberate growth promoting substances such as auxin and amino acids, which stimulates growth of rice plants, supply additional humus to the soil, improve aeration in rice roots and reduce sulphide injury. It is applied at 10 kg per hectare after 10 d of transplanting. The field is kept waterlogged for a couple of days. Sun dried algae material can be stored for long and used in the field. The amount of N fixed by BGA is 15 – 45 kg per hectare. Application of 3 – 4 t of FYM per hectare is recommended for higher N fixation.

Multiplication of BGA

- **Trough Method:** Troughs of 2 m × 1 m × 22 cm capacity each of galvanized iron sheet are filled with 5 – 10 cm water and placed under sunlight. A mixture of 8-10 kg soil and 200 g super phosphate + lime at 30 g is sprinkled uniformly on the water. Then mother culture at 250 g per tray is added and left under sun. In 7 – 10 d a thick mat of BGA appears. Water is allowed to evaporate and dried algae flakes are collected and used. Usually 5 – 7 kg BGA obtained from one tray.
- **Pit Method:** Shallow pits of similar dimension as trough method are made, lined with polythene sheet to prevent drainage. Other procedure remains the same.

- **Field Method:** Selected field is ploughed and levelled and beds are formed. Water is allowed to stand for 5 cm depth. Application of 12 kg SSP (in organic farming use rock phosphate) and 200 g lime ($40m^2$) followed by composite algae culture at 5 kg is required. In 10 – 20 days a good mat covers the bed and field is allowed to dry. Then flakes are collected. During summer months, a quantity of 30 – 40 kg per 40 m^2 can be obtained.

(iv) Azolla – Azolla is a free floating fresh water fern. *Azolla pinnata* is the most common species occurring in India. It fixes atmospheric N due to *Anabaena* species of BGA present in lobes of azolla leaves. A thick mat of azolla supplies 30-40 kg N per hectare. Normal temperature for growth is 20 to 30° C. It thrives better in monsoon season with frequent rains and cloudiness. Suitable soil pH is 5.5 – 7.0.

- **Nursery:** Field should be well prepared and levelled uniformally, divided into beds of 20 × 2 m size by providing suitable bunds and irrigation channels. Beds may be lined with polythene sheet. Water is maintained to a depth of 10 cm. Ten kilogram of fresh cattle dung, mixed in 20 litres of water is sprinkled in each plot and azolla inoculum at 8 kg is introduced to each bed. Single super phosphate is applied at 100 g per bed in three splits at 4 d interval as top dressing (not in organic farming). After 15 d it is harvested and applied in the main field as source of primary inoculum. From one harvest, 40 – 50 kg fresh azolla is obtained from each bed. Four nursery beds each of 40 m^2 are required for raising azolla for inoculation of one hectare main field.
- **Field application of azolla:** Azolla is applied to the main field as a green manure crop or as a dual crop. As green manure crop, azolla is allowed to grow on flooded fields for 2-3 wk before transplanting. Later water is drained and azolla is incorporated by ploughing. As a dual crop, 1,000 to 5,000 kg per hectare of azolla is applied to the soil one week after translating. When thick mat formed, it is incorporated by trampling. The left over azolla develops again which is trampled in as second crop.

4.3.3 Other Bio-inoculants

(i) **Phosphate Solubilizing Micro-organisms (PSM)**: Only 25 – 30 per cent of applied phosphorus become available to the crop and remaining part is fixed. Fixed form of soil phosphate is solubilized by a group of micro organisms called PSM. Some micro-organisms like *Bacillus megatherium, pseudomonas straita* (bacteria), *Aspergillus acva, Trichoderma spp.* (fungi) are capable of solubilizing phosphates. The inoculum of these microbes are applied as that of other inocculants.

(ii) **Vesicular Arbuscular Mycorrhiza (VAM):** It assimilate P, Zn and other nutrients from the soil for their own need and in addition translate them in different forms to the host roots. Important crops associated with this fungi are wheat, maize, millets, beans, potato, soybean, tomato, grapes, apple, banana, sugarcane, castor, tobacco, tea, coffee, cocoa and rubber. VAM directly translocate the nutrients like P, Zn, Cu, K, Al, Mn and Mg from the soil to the root cortex and increases the growth of associated plants by producing auxins, antibiotics, etc. Inoculation methods for VAM are given as under –

- **Transplanted crops:** The seedlings are raised in the soil supplied with VAM in small nursery beds or containers and planted out when mycorrhizal inoculation is well established.
- **Direct sown crops:** Seeds are coated with an adhesive methyl cellulose to which inoculum is expected to stick and subsequently infect the emerging radicles. Furrow inoculation is effective method of inoculating field sown crops where inoculum placed under or beside seeds sown in furrows.

(iii) **Saprophytes:** Micro-organisms that are capable of decomposing organic matter at faster rate for release of nutrients are *Asperagillus, Penicillium, Trichoderma,* are cellulolytic fungi which breakdown cellulose of plant material. The natural process of decomposing is accelerated and compositing time is reduced by 4 – 6 wk.

4.4 SOME IMPORTANT FORMULATIONS FOR SOIL ENRICHMENT

Preparation of liquid manures

Many variants of liquid manures are being used by farmers of different states. Few important and widely used formulations are given below:

(i) **Sanjivak:** Mix 100 kg cow dung, 100 litre cow urine and 500 g jaggery in 300 litre of water in a 500 litre closed drum. Ferment for 10 days. Dilute with 20 times water and sprinkle in one acre either as soil spray or along with irrigation water.

(ii) **Jivamrut:** Mix cow dung 10 kg, cow urine 10 litre, Jaggery 2 kg, any pulse grain flour 2 kg and live forest soil 1 kg in 200 litre water. Ferment for 5 to 7 d. Stir the solution regularly three times a day. Use in one acre with irrigation water.

(iii) **Amritpani:** Mix 10 kg cow dung with 500 g honey and mix thoroughly to form a creamy paste. Add 250 g of cow *desi ghee* and mix at high speed. Dilute with 200 litre water. Sprinkle this suspension in one acre over soil or with irrigation water. After 30 d apply second dose in between the row of plants or through irrigation water.

(iv) **Panchgavya:** Mix fresh cow dung 5 kg, cow urine 3 litre, cow milk 2 litre, curd 2 litre, cow butter oil 1 kg and ferment for 7 days with twice stirring per day. Dilute 3 litre of *panchgavya* in 100 litre water and spray over soil. Twenty litre of panchgavya is needed per acre for soil application along with irrigation water.

(v) **Enriched Panchgavya (or Dashagavya):** Ingredients - cow dung 5 kg, cow urine 3 litre, cow milk 2 litre, curd 2 litre, cow *deshi ghee* 1 kg, sugarcane juice 3 litre, tender coconut water 3 litre, banana paste of 12 fruits and toddy or grape juice 2 litre. Mix cow dung and *ghee* in a container and ferment for 3 d with intermittent stirring. Add rest of the ingredients on the fourth day and ferment for 15 d with stirring twice daily. The formulation will be ready in 18 d. Sugarcane juice can be replaced with 500 g jaggery in 3 litre water. In case of non-availability of toddy or grape juice 100g yeast powder mixed with 100 g jaggery and 2 litre of warm water can also be used. For foliar spray 3-4 litre *panchgavya* is diluted with 100 litre water. For soil application 20 litre *panchagavya* is sufficient for one hectare. It can also be used for seed treatment.

Chapter 5

Insect Pests and Disease Management in Organic Farming

Global concern about pesticide use, their residues in soil and plant systems as well as their hazardous effect on ecology and environment have diverted the attention of scientists to find out an eco-friendly approach of insect pests and disease management. Organic farming is one among broad–spectrum of production methods which is supportive of the environment. However, insect pests pose a major challenge in organic production system since genetically modified crops and synthetic pesticides are not permitted for use in organic production system. Therefore, it has been realized to formulate eco-friendly strategies for the integrated management of insect pests and diseases in organic farming.

In organic farming there is need to follow a pest management system which includes the integration of all known biological, behavioral, cultural methods, or organically accepted chemical alternatives (pyrathium, botanical toxins, horticultural oils & fatty acids). The integrated pest management also should be knowledge intensive and farmer based management approach that encourages natural control of pest populations by anticipating pest problems and preventing pests from reaching economically damaging levels. Successful pest management depends on integration of a number of control strategies. Some strategies will target insect pests and diseases separately and others will target them together. Different methods available for management/control of insect pests and diseases in organic farming is discussed as under:

5.1 BIOLOGICAL METHODS

The successful control of pest and diseases by means of another organisms that is encouraged and disseminated by man is known as bio-control. This means use of any biotic agents for minimizing the pest population either directly or indirectly. Biological control of insect pests is gaining recognition as an essential component of successful integrated pest management (IPM) in organic farming. Classical biological control involves deliberate and natural establishment of natural enemies in areas where they did not previously occur. In addition do deliberate introduction of bio-control agents, proper attention needs to be given for conservation and augmentation of natural enemies that already exist in an area. Biological pest control is an ideal alternative to control pests, using natural enemies. These are highly specific, and safe to non-target species, beneficial insects, animals and man. Moreover, these are less expensive and have long term control, causes no pollution of pest / crop agroecosystem. The principal ways for bio-control are as follows:

(i) Collection of bio-control agents viz., parasitoids and predators.

(ii) Isolation of antagonizing microbial organisms containing bacteria, fungi, viruses, nematodes, protozoa, etc. which are known as **microbial pesticides**.

(iii) Using plant based materials like neem which are known as **botanical pesticides,** used for controlling pests.

5.1.1 Bio-control Agents

5.1.1.1 *Parasitoids*

The parasitoids like *Trichogramma* spp., *Bracon brevicornis, Telenomus remus, Abenteles angaleti,* etc. have potential to control pests like sugarcane borer, maize and sorghum borer, paddy borer and cotton boll worms. Parasitoid immatures develop on or inside a host, killing it as they mature. They emerge as adults and continue the cycle. *Trichogramma* spp. is most effective parasitoids as biotic agent mainly for borers.

5.1.1.2 *Predators*

Predators (Chrysopids and Coccinellids) may voraciously feed on aphids, mealy-bugs and other insects attacking cotton, tobacco, sunflower, mustard, groundnut, citrus, guava, grape, etc. (Figs. 5.1 *a* & *b*). A predator consumes many prey during its lifetime. *Chrysoperla carnea* is most important insect predator for target pests like aphids, whitefly, mites, boll worms, caterpillar, etc. Other important predators *are Curinous coeruleus, Adonia variegate,* etc.

(*a*) *Coccinelle*

(*b*) *Chrysopids*

Fig. 5.1. Photographs of Predators

5.1.2 Microbial Pesticides

5.1.2.1 *Bacterial Pesticides*

Bacillus thuringiensis (Bt) is most commonly used bacterial pesticide globally. It is mainly a pathogen of lepidopterous pests, which includes American boll worm in cotton and stem borer in rice. The toxicity of Bt strains have been attributed to a glycoprotein, protoxin, often referred as delta endotoxin which is produced during sporulation/ fermentation. It is a crystalline product. When Bt spores are ingested by the larvae, they are activated within the gut to produce toxins, which damages the mid gut of pest eventually killing it. *Pseudomonas and fluorescence* (produce HCN) are also used as bacterial pesticides. Others are *Bacillus subtilis, Azotobacter*, etc.

5.1.2.2 *Fungal Pesticides*

Trichoderma is commonly used as fungal pesticide. It is an antagonistic fungi for controlling a number of seed borne, seedling diseases of plants like root rot, charcoal rot, *Fusarium* wilt diseases on pulses, oilseeds, sorghum and cotton. *Trichoderma* is also effective in horticultural crops like carnations, petunia, tomato, pepper, cucumber and other vegetable crops. It is applied as seed treatment at the rate of 4-8 g per kg of seed and also as soil incorporation along with farmyard manure or neem cake. *Trichoderma viride* and *T. harzianum* are commonly used for this purpose.

5.1.2.3 *Viral Pesticides*

Baculoviruses are commonly used as viral pesticides. These baculoviruses are pathogenic to anthropods only and are considered highly safe to non–target organisms including man and the environment. It is estimated that these viruses can be used against nearly 30 per cent of all major food and fibre crop pests. Baculoviruses are of two types - viz., (i) Granulosis Virus

(GV) in which virus particles are packed singly and (ii) Nuclear Polyhedrosis Virus (NPV) where the virus particles are aggregated.

Two NPVs are particularly important for India. These are very effective against *Heliothis armigera and H. spodoptera* (*Heliothis* NPV and *Spodoptera litura* NPV), which are common pests of cotton, maize, chickpea, groundnut, tobacco, etc. NPVs can be used against cotton boll worm and tobacco bud worm at the rate 140-250 g per hectare. The wettable of NPV developed by Sandoz (India) Ltd. contains atleast 4 million polyhedral inclusion bodies (PIB) of *Heliothis* NPV per g of product. *Chilo infuscatellus* GV belongs to the group Granulosis Virus which is similar in the antagonistic characteristics to other baculovirses; it is extremely target specific. It is active against *Chilo infuscatellus* (sugarcane borer), an important pest of sugarcane.

The major problems with Bt and viral preparations are that they are systemic in nature, selective, less persistent and do not act on contact. However, being specific, bio-degradable, cost effective and of increased safety to environment, such product hold much promise. Moreover, with appropriate bio-technological and genetical techniques such as mutation, gene cloning and process optimization, it should be possible to develop useful new strains. With the help of these modern techniques, it should also be possible to improve, stabilize and deliver baculoviruses and Bt toxins to control wide varieties of insect species.

The accepted limit of bio-pesticides are

- Colony Forming Unit (CFU) = 2×10^8 per ml or per g
- Contamination level = 1×10^5 per ml or per g
- Moisture (%) is more than 8
- Shelf – life is 4–6 months

5.1.3 Botanical Pesticides

Botanical pesticides like *Azadirachta indica, Annona squamosa, Ryania specicosa, Pongamia glabra, Acorus calamus, Anethor sowa and Madhula longifoila* are being used to control pests. Among all botanical pesticides *Azadirachta indica (neem)* is found most effective bio pesticide as it contains several chemicals, including "azadirachtin" which affects the reproductive and digestive process of a number of important pests.

Neem has been found to be effective in the management of approximately 200 insects, pests and nematodes. Neem is very effective against grasshoppers, leaf hoppers, plant hoppers, aphids, jassids, and moth caterpillars. Neem extracts, are also very effective against beetle larvae, butterfly, moth and caterpillars such as Mexican bean beetle, Colorado potato beetle and diamond back moth. Neem is very effective against grasshoppers, leaf minor and leaf hoppers such as variegated grasshoppers, green rice leaf hopper and cotton jassids. Neem is fairly good in managing beetles, aphids, white flies, mealy bug, scale insects, adult bugs, fruit maggots and spider mites.

Advantages of biological control methods

(i) Bio-control is exercised in a wide range of area and is safe for human and animal health.

(ii) Application of biotic agents is easy and possible even in inaccessible areas like dense forests, tall trees, ponds, rivers, lakes and ravines.

(iii) The biotic agents survive in nature till the pest is prevalent and self perpetuating in nature.

(iv) Farmer does not require any special treatment procedure/ application equipments viz., sprayer, duster, etc. except for microbial preparations which may be mixed and sprayed along with other chemicals and fertilizers.

(v) No waiting period for harvesting of a crop after treatment with biotic agents.

(vi) Biological agents, viz., baculoviruses, parasitoids and predators may be multiplied at farmer's level as these do not require sophisticated equipments.

5.1.4 Biopesticides and IPM Products

Biopesticides and IPM products used in different crops to control insect pests and diseases are summarized in table 5.1 as under.

Table 5.1: Bio-pesticides and IPM Products Used in Different Crops.

Crops	Insect pests and diseases	Biopesticides & IPM products
Cotton	Bollworms	Traps, Lures, BT, NPV, *Trichogramma*
	Whitefly, Jassides, thrips	Neem 1500 ppm
	Mites	*Chrysoperla, Verticillium, Baeluveria*
	Wilts and leaf spots	*Trichoderma, Pseudomonas*
Rice	Yellow stem borer, leaf folder	Traps, Lures, BT, *Trichogramma*
	Hoppers, Sheath blight, leaf spots	Neem 1500 ppm, *Baeluveria*
Chickpea, Pigeonpea, Greengram, blackgram	Pod borer/ cut worms	Traps, Lures, BT, NPV, *Trichogramma*
	Wilts	*Trichoderma, Pseudomonas*
Sugarcane	Borers	*Trichogramma*
Soyabean	Caterpillars	Traps, Lures, BT, NPV, *Trichogramma*
Mustard	Aphids	Neem 1500 ppm, *Chrysoperla*
	White rust & leaf spots	*Trichoderma, Pseudomonas*
Potato	Cutworms	Traps, Lures, BT
	Early blight, late blight	*Trichoderma, Pseudomonas*

5.1.5 Some Other Pest Control Formulations

Many organic farmers and NGOs have developed large number of innovative formulations which are effectively used for control of various pests. Although none of these formulations have been subjected to scientific validation but their wider acceptance by farmers speak of their usefulness. Farmers can try these formulations, as they can be prepared on their own farm without the need of any ingredients from the market. Some of the popular formulations are listed below:

(i) **Cow Urine:** Cow urine diluted with water in ratio of 1: 20 and used as foliar spray is not only effective in the management of pathogens and insects, but also acts as effective growth promoter for the crop.

(ii) **Fermented Curd Water:** In some parts of central India fermented curd water (butter milk or *chaach*) is also being used for the management of white fly, jassids. aphids, etc.

(iii) **Dashparni Extract:** Crush neem leaves 5 kg, *Vitex negundo* leaves 2 kg, *Aristolochia* leaves 2 kg, papaya leaves 2 kg, *Tinospora cordifolia* leaves 2 kg, *Annona squamosa* (custard apple) leaves 2 kg, *Pongamia pinnata* (karanj) leaves 2 kg, *Ricinus communis* (castor) leaves 2 kg, *Nerium indicum* 2 kg, *Calotropis procera* leaves 2 kg, green chilli paste 2 kg, garlic paste 250 g, cow dung 3 kg and cow urine 5 litre in 200 litre water ferment for one month. Shake regularly three times a day. Extract after crushing and filtering. The extract can be stored up to 6 mo and is sufficient for one acre.

(iv) **Neem-Cow Urine Extract:** Crush 5 kg neem leaves in water, add 5 litre cow urine and 2 kg cow dung, ferment for 24 h with intermittent stirring, filter, squeeze the extract and dilute to 100 litre, use as foliar spray over one acre. Useful against sucking pests and mealy bugs.

(v) **Mixed Leaves Extract:** Crush 3 kg neem leaves in 10 litre cow urine. Crush 2 kg custard apple leaves, 2 kg papaya leaves, 2 kg pomegranate leaves, 2 kg guava leaves in water. Mix the two and boil 5 times at some interval till it becomes half. Keep for 24 h, then filter, squeeze the extract. This can be stored in bottles for 6 mo. Dilute 2-2.5 litres of this extract to 100 litre for one acre. Useful against sucking pests, pod/fruit borers.

(vi) **Chilli-garlic Extract:** Crush 1 kg *Ipomea* (besharam) leaves, 500 g hot chilli, 500 g garlic and 5 kg neem leaves in 10 litre cow urine. Boil the suspension 5 times till it becomes half. Filter, squeeze the extract and store in glass or plastic bottles.

2-3 litre extract diluted to 100 litre is used for one acre. Useful against leaf roller, stem/fruit/pod borers.

(vii) **Broad Spectrum Formulation - 1:** In a copper container mix 3 kg fresh crushed neem leaves and one kg neem seed kernel powder with 10 litre of cow urine. Seal the container and allow the suspension to ferment for 10 d. After 10 d boil the suspension, till the volume is reduced to half. Ground 500 g green chillies in one litre of water and keep overnight. In another container crush 250 g of garlic in water and keep overnight. Next day thoroughly mix the boiled, chilli, and garlic extract and filter. This is a broad spectrum pesticide and can be used on all crops against wide variety of insects. Use 250 ml of this concentrate in 15 litre of water for spray.

(viii) **Broad Spectrum Formulation - 2:** Suspend 5 kg neem seed kernel powder, one kg *karanj* seed powder, 5 kg chopped leaves of *Ipomea* spp. (besharam) and 5 kg chopped neem leaves in a 20 litre drum. Add 10-12 litre of cow urine and fill the drum with water to make 150 litre. Seal the drum and allow it to ferment for 8-10 d. After 8 d mix the contents and distil it in a distiller. Distillate will act as a good pesticide and growth promoter. Distillate obtained from 150 litre liquid will be sufficient for one acre. Dilute in appropriate proportion and use as foliar spray. Distillate can be kept for few mo without any loss in characteristics.

5.2 BEHAVIOURAL METHODS

5.2.1 Pheromone Traps

Pheromone is ectohormons secreted by an organism, which elicit behavioral responses from other members/ sex of its own species. These are extremely selective, non-toxic, highly bio-degradable and effective at low application rates. Synthetic sex pheromones are commercially available and are used for surveillance, monitoring and control of many *Lepidopterous* pests such as spotted bollworm, tobacco catterpiller, potato tuber moth, diamond back moth and leaf folder, etc. Pheromone traps baited with the sex attractant (7z, 11 z) – 7, 11 hexadecadienal can catch on an average of 103.09 adults per day over a period of 6 d.

Fig. 5.3. Phtographs of Pheromone Traps

5.2.2 Fairomones

These are volatile compounds that evoke behavioral response adaptively favourable to the receiver. Fairomones are released either by host plant or by the host insects while former issued by the pest and natural enemies to locate their habitats, the later is used for prey finding and parasitisation preying. Fairomones from host plant can be effectively used to mass trap pest species as well as for monitoring. Use of fairomonal compounds to increase the efficiency of predators *Chrysoperla carnea* and the egg parasitoid *Trichogramma chilonis* had been successfully demonstrated.

5.3 CULTURAL METHODS

Cultural control is purposeful manipulation of the environment to make it less favourable to their natural enemies. The important cultural methods to control pests and diseases are as follows:

5.3.1 Selection of Varieties

Varieties/cultivars with yield potential and quality without resistance / tolerance to pests and diseases are the main causes of frequent epidemics and multiplication of insect pests and diseases. A large number of cultivars/varieties resistant / tolerant to different pests and diseases have been developed to suit to different agro-ecosystems.

The selection of insect pests and disease resistant cultivars can be a useful tool, but under no circumstances can genetically modified varieties be used in organic farming system. Wheat varieties with solid stems are more resistant to wheat stem sawfly than hollow stemmed varieties.

Certain species may avoid diseases such as *Fusarium* head blight, but often agronomic factors, such as time of seeding or choosing winter versus spring wheat, have more of an influence on the incidence of diseases. Plants also vary in their degree of attractiveness to insects, diseases and vectors transmitting diseases. Factors such as leaf and stem toughness, pubescence, nutrient content, plant architecture, growth habit and differences in maturity between crops and varieties can influence pest growth, reproduction and host preference. For example, early-maturing crop varieties may be less attractive to migrating populations of grasshoppers late in the season compared to later-maturing varieties.

5.3.2 Seed Quality

The use of high - quality seed is especially important in preventing diseases. The seed should be free of smut, ergot bodies or other sclerotia, and free of kernels showing symptoms of *Fusarium* head blight infection. Relatively few diseases are exclusively seed-borne, and it is more common for pathogens to be transmitted from soil, stubble, or wind, as well as with the seed. Planting physically sound seed is also important. In crop such as flax, rye and pulses, a crack in the seed coat may serve as an entry point for soil-borne micro-organisms.

5.3.3 Tillage

Ploughing and intercultivation brings about unfavourable conditions for multiplication of insect pests and diseases. Quiescent stages (pupal) of harmful organisms will be expected to dehydration or to predation by birds and other stages may be mechanically damaged or furred deep in. In view of this tillage can be properly timed before seeding, after harvesting and during summer fallow to reduce population of insect pests such as cutworms and grass hoppers that spend part of their life cycles in soil or stubbles. Tillage can help to starve insects in spring or during fallow, prevents adults from laying eggs in the soil and expose overwintering to predators and inclement weather.

5.3.4 Sowing Time

Planting should be scheduled in such a way that the most susceptible time of plant growth does not correspond to the peak insect pests cycles. Early seeding reduces crop damage caused by grasshoppers, aphids in cereal crops, wheat midge in spring wheat, barley, yellow dwarf virus in barley and wheat and, powdery mildew in peas. Delayed seeding can be effective in avoiding wireworms and cutworms in cereal crops, barley thrips, and wheat streak mosaic virus in winter wheat. Maize sown late suffers little borer damage as by then egg parasitoid *Trichogramma* is able to keep down the population of pest. However, delaying seeding too long can reduce the crop productivity. Rice and sorghum suffer less from borer and shoot fly attack, respectively if planted early (early June).

5.3.5 Depth of Seeding

Optimum seeding depth is also important. Deep seeding in cold soils may result in seedling blights and damping-off, especially in pulses and small seeded crops. Seeding depth should generally be no deeper than required for quick germination and even emergence. If the soil is loose before seeding, a packing operation will firm up the soil and bring moisture closer to the surface. For most crops, seeding should ideally be done when the soil is warm enough for rapid germination. Seeds that remain ungerminated in cool soil are more susceptible to damage by insects such as wireworms.

5.3.6 Plant Population

Plant population per unit area influence crop micro–climate. Dense plant canopy leads to high humidity build up congenial for insect pests and diseases multiplication. Therefore, optimum plant population as per recommendation for individual crop may be maintained to minimize the humidity build up within the plant canopy and their by avoiding multiplication/ infestation of insect pests and diseases.

5.3.7 Crop Rotation

Pest problems can be reduced to a large extent by carefully rotating a crop belonging to one family with another from different groups. Rotating groundnut with maize will reduce white grub attack. Rotating pigeonpea or chickpea with a non-leguminous crop is a good practice for managing *Fusarium* wilt and nematode problems. Diverse rotations are particularly effective in regulating flea beetles, butter fly, wheat midge, wheat stem maggot and wheat stem saw fly. Rotations are also effective in controlling soil and stubble borne diseases. The success of rotations in preventing diseases depends on many factors, including the ability of a pathogen to survive without its host and the pathogenic host range. Those with wide range of hosts will be controlled less successfully.

5.3.8 Intercropping

Intercropping with diverse plant geometry breaks the pattern of standard monocropping and limits infestation from insects. Diverse planting not only obstructs adults from egg laying but also the release of volatile allelo – chemicals from particular crops. These effects do not provide suitable conditions for the development of a micro climate favouring the multiplication of a particular type of pest. For example cabbage along with carrot or tomato, is an important intercrop combination to effectively manage diamond back moth.

5.3.9 Trap Crops

It is an effective way to minimize most pest problems. Raising trap crops and inter and / or border crop is an important cropping system approach. The trap crop must be distinctly attractive to the pest when compared with the main crop. It provides protection either by preventing the pest from reaching the main crop or restricting them to a certain part of the field where they can be economically destroyed. Mustard along with cabbage is a trap crop for the control of the diamond back moth, aphids and leaf webber. African marigold is a good trap crop for American boll worm, besides it also attracts the adults of the leaf minor to lay eggs on its leaves. Maize plants can be a trap crop to attract fruit fly in vegetable cultivation where the fruit fly is a major problem.

5.3.10 Trap Strips

Seeding trap strips around the edge of a cropped field or along a fence row helps to lure insect pests to a specific area where they can be managed more easily. For example, planting brome grass near a wheat field attracts wheat stem saw flies and their native parasites away from the wheat crop. Similarly, a trap strip of potatoes planted much earlier than the main crop would attract colorado potato beetles to the area. The strip could be worked under along with the adult beetles, eggs and larvae before the second generation of beetles spreads to the main crop.

Generally, the insect pests in the trap strips are controlled by mowing or cultivating the strip, or by applying an acceptable organic product, such as *Bacillus thuringiensis*. Trap strips can also act as a barrier to protect the crop field. Producers have found that planting yellow sweet clover or sirius field peas repels grasshoppers and prevents them from damaging crops.

5.3.11 Water Management

Irrigation can reduce the soil inhabiting pests by suffocation or exposing them to soil surface to be preyed upon by birds. Irrigating potato crop at tuber formation can minimize potato scab. Anthracnose of beans, early blight and charcoal rot of potato can be checked by furrow irrigation than sprinkler irrigation.

5.3.12 Balanced Nutrition

Balanced soil nutrition is essential for crop quality, yield and moisture use efficiency. The application of nutrients should be based on a sound soil testing program, accompanied by plant tissue analysis when diagnosing problems. High levels of nitrogen can occur after a high-nitrogen plow down, such as sweet clover. This results in lush leaf tissue and a dense plant canopy that provides an ideal environment for plant pathogens. However, a lush crop may also help in disperse the damage by a given number of insects, so accurate observations by the producer are necessary at all times. In contrast, inadequate soil phosphorous can predispose to certain root diseases. Low levels of nitrogen can reduce the incidence of insect outbreaks. A shortage of micronutrients such as zinc or copper can result in disease-like symptoms on crops, while too much of any one micronutrient may be toxic. The addition of composted livestock manure improves soil quality, including increasing the population of soil micro-organisms that compete with soil-borne plant pathogens.

5.3.13 Sanitation

Sanitation includes the removal or destruction of breeding refuges and over wintering sites of pests. Removing diseased plants (rouging) or plant parts is important in disease management. To prevent introduction of inoculum, all means of carrying the pathogen should be looked into e.g., seed, water, plant debris, farmyard manure, etc. Pest population may often be effectively suppressed by destruction of their alternate plant hosts, many may infest weeds. Destruction of infested young shoots of sugarcane and stubbles after harvest helps a great deal in keeping the incidence of sugarcane borer at very low level.

5.3.14 Forecasting

Producers should pay attention to the forecasts for various insect pests and disease infestations for each crop year. Maps of these forecasts are usually available for many of the major destructive insects such as grasshoppers and wheat midge, as well as some diseases. Agro-meteorological warning and forecast can help in this way.

5.3.15 Bird Perches

Bird perches are resting places for predatory birds to rest and to look for preys; such as insect pests of cotton, peanuts, and cowpeas. Predatory birds prefer to look for prey in field crops where they have places to rest. To make bird perches, use bamboo or wooden poles or tree branches. Erect either of these at regular intervals in the field. To have live bird perches within the field, plant *Setaria* spp. (foxtail). These plants are found to be attractive to predatory birds. The birds feed on their seeds. In cotton field, plant *Setaria* spp. in every 9th or 10th row of cotton. Once the birds are on the field, they prey on cotton bollworms and other insects.

Chapter 6

Weed Management in Organic Farming

Weeds compete with crop plants for mineral nutrients, moisture, light, space, etc, which adversely affect the growth and yield of the crop. Of the total loss of agriculture produce from various pests, weeds accounts for 45 per cent, insect pests 30 per cent, diseases 20 per cent and other pests 5 per cent. Data available on losses caused by uncontrolled growth of weeds show yield reduction of 35 per cent to 90 per cent in different crops and cropping situations. The average losses in crop yield on the country basis have been estimated at 15-30 per cent in wheat, 30-35 per cent in rice and 15-85 per cent each in maize, sorghum, pulses and oilseeds. But as farmers adopt some kind of weeding in field, it still leaves us with a conservative estimate of atleast 10 per cent reduction in crop yields. This reveals that weeds are the most costly category of agricultural pests causes maximum yield losses and added labour costs than either insect pests or crop diseases.

The critical period of weed crop competition is of paramount importance in life cycle of the crop from weed control point of view. The critical period of weed crop competition is different for different crops. Generally a crop of 90-100 d duration, the first 30-35 days after sowing (DAS) should be maintained weed free for optimum yield. The critical period of weed crop competition for maize and sorghum 15-45 d DAS, rapeseed and mustard 15-40 DAS, soybean 20-45 DAS, *kharif* pulses 15-30 DAS, pigeonpea 15-60 DAS, groundnut 30-50 DAS, wheat 30-45 DAS, direct seeded rice 15-45 DAS, transplanted rice 30-45 DAS, chickpea 30-60 DAS, and sugarcane 30-120 DAS.

For management of weed in organic systems, they rely only on non-chemical or ecological approaches. A farmer who manages weeds organically must be intimately familiar with the type of weeds and their growth habits to determine which control method to be employed. Its management begins with careful planning of the cropping system to minimize weed problems, and seeks to utilize biological and ecological processes in the field and throughout the farm ecosystems to give crops the advantage over weeds. Important weed control methods or weed management practices followed in relation to organic farming includes preventive measures, good crop husbandary/ cultural methods, physical or mechanical methods and biological methods. A brief description of these weed control methods is given as under.

6.1 WEED PREVENTION

Weed prevention comprises all measures which deny the entry and establishment of weeds in an area. In other words, all practices that help to discourage the weeds from becoming a problem over time, for the subject of weed prevention. Thus, any physical or chemical method employed with the main objective of not allowing the weeds to set viable seeds, is to be considered as a part of weed prevention. Weed prevention is very important as it reduces the farmers' efforts in controlling weeds later by physical or other methods of weed control and thereby saving lot of labour and cost involved in weed management. The weed preventive measures are extremely important against obnoxious weeds, as their control/ eradication is extremely difficult if once established in the crop field/non-cropped areas. Important preventive measures to be adopted are given as under:

(i) Sowing of clean seeds, free from weed seeds.
(ii) Avoiding feed material, containing weed seeds to the farm animals.
(iii) Avoid adding weeds to the manure pit. Always use well decomposed FYM / compost.
(iv) Use clean farm machineries free from weed seeds / propagules of weeds. Also clean farm machinery thoroughly especially seed drills before moving it from one field to another one.
(v) Destruction of weeds emerged on bunds of fields, irrigation channels and non-cropped areas nearby the crop field.

(vi) Transplanting of seedlings free from germinating weeds.

(vii) Prohibiting the entry of livestock / cattle in infested areas with noxious weeds.

(viii) Destruction of weeds before flowering from roads, paths, barren lands and waste lands.

(ix) No shifting of soils from weed infested areas to another fields.

(x) Use visilance. Inspect the farm frequently for any strange looking weeds. Destroy such patches of new emerged weeds by digging deep and burning it along with its roots.

6.2 WEED CONTROL METHODS

Weed control in crop fields is the process of limiting any given weed infestation to the extent that it permits economic crop production. The objective here is to limit the growth of unwanted plants without any attempt to eliminate them from the scene. The extent to which a given weed growth should be limited will depend upon the cost involved in operation and the benefits anticipated from the operation. The methods employed in organic farming for weed management are be dealt as under:

6.2.1 Good Crop Husbandry / Cultural Methods

(i) **Proper Planting Methods:** Any planting method that leaves the soil surface rough and dry will discourage early weed growth. For example, planting of winter grains in seed beds prepared after a presowing irrigation, it leaves top 3 to 5 cm soil above the crop seeds in rough and dry tilth. Such a physical condition of soil defers the germination of the weeds. Likewise in rainy season also farmer should complete their seed bed preparation and sowing of crops with minimum time lapse. Plough–planting and other minimum tillage planting methods have often proved very useful in minimizing early weed problems. In summer, furrow planting of crops is very useful method for reducing the weed problem.

(ii) **Proper Planting Time:** Manipulating the sowing time of a particular crop earlier or later compared to conventional sowing time helps to minimize the weed infestation. For example growing of maize, cotton 15 days before break of monsoon with the help of presowing irrigation and late planting of suitable varieties in winter season reduces weed infestation.

(iii) **Proper Crop Stand:** Gappy and under populated crops are prone to heavy weed infestation which become difficult to control later. Therefore, practices like selection of proper seed, right method of sowing, adequate seed rate, protection of seed from insects, pests and diseases are very important to obtain proper and uniform crop stand capable of offering initial competition to the young weeds.

(iv) **Selective Crop Stimulation:** Application of fertilizer as basal and top dressing to the crop helps in selective stimulation of crop plants which can withstand competition from weeds much better. If the soil is problematic, not suited much to the crop, use of proper soil amendments should be considered necessary to boost crop growth and, thus indirectly suppress the weeds.

(v) **Crop Rotation:** Parasitic weeds, as well as the crop associated weeds, can be discouraged by adopting well conceived crop rotations. For example a grassy weed like *Phalaris minor* can be easily discouraged by replacing wheat with a forage crop in alternate years.

(vi) **Summer Fallowing:** In major parts of India there is clear cut solar energy rich dry period of summer available, which should be utilized for desiccation of rhizomes, tubers and roots of perennial weeds to death. Initial tillage of the field in summer for this purpose should encourage clod formation. These clods, which embody the weed propagules, upon drying, desiccate the weed propagules within, much better than a pulverized soil. Subsequent tillage operations should break these clods into smaller units to further expose the shrivelled weeds to the hot sun.

(vii) **Reduction in Area under Bunds (levees):** Farmers often ignores the weeds on bunds and channels of crop fields and this becomes the important source of weed seed for propagation. Therefore, it is very important to reduce the area under bunds and channels.

(viii) **Stale Seed Bed:** A stale seedbed is one where initial one or two flushes of weeds are destroyed before planting of a crop. This is achieved by soaking a well prepared field with either irrigation or rain and allowing the seeds to germinate. At this stage a shallow tillage may be followed to destroy the dense flush of young weed seedlings. This may be followed immediately by sowing a desired crop.

(ix) **Solarization:** This is another method of utilization of solar energy for the desiccation of weeds. In this method the soil temperature is further raised by 5-10^0 C by covering a pre-soaked fallow field with transparent plastic. The plastic sheet checks the long wave back radiation from the soil and also prevents loss of energy by hindering moisture evaporation. It is very good practice for high value crops as initial cost is very high.

(x) **Intercropping:** Several times intercropping helps in the suppression of atleast the secondary growth of weeds that occurs after the intercrop has fully covered the ground. The weeds occurring before the intercrop has shaded the inter row spaces should be controlled by mechanical means. Cowpea, blackgram and greengram are common intercrops in India used to increase the total crop productivity and suppress the secondary growth of weeds.

(xi) **Minimum Tillage:** It is now well accepted that the primary objective of any tillage programme on the farm is to control weeds. If the weeds were effectively controlled by some other method, the minimum tillage and zero tillage crop cultivation practices can be adopted. In India, however, the preplant tillage operations used for seedbed preparation are still the most practical means to obtain initially weed – free fields for sowing of the crops. An ideal physical soil condition for sowing of seed row crops is to obtain a firm tilth in the wide rows and a loose tilth in the inter rows. This will ensure rapid crop seed germination but slow weed emergence in the inter-row spaces.

(xii) **Flooding and Drainage:** Submergence of weeds in water for some time destroys the weeds, when sufficient water is available and crop like rice can sustain under flooding conditions. In variance with flooding, drainage is used for controlling aquatic and semi – aquatic weeds in rice fields, channels, canals, and ponds. In rice fields, where both terrestrial and aquatic weeds may be common, a judicious combination of the two can be practiced.

6.2.2 Physical or Mechanical Methods

The weeds are controlled with the help of suitable farm implements / machineries operated by manual or animal or mechanical power depending on the type and intensity of weed flora. Following mechanical / physical methods of weed control are commonly practiced.

(i) **Tillage:** The primary function of any tillage operation is to control weeds. Preplant tillage is constituted of both primary (using disc plough, MB plough, etc.) and secondary tillage (using cultivator, disc harrow, etc.). The primary tillage help to control weeds by burying the existing weeds, bringing the weed seeds to soil surface for germination followed by destructions by suitable secondary tillage. Post plant tillage operations done by different implements like spike tooth harrow, rotary hoe cultivator, wheel-hoe, blade harrow, etc. are found very useful to control weeds at initial stage of crop growth. These implements are very effective and economic as compared to other methods like hand pulling, hand hoeing, etc. These implements are also very useful in breaking the soil crust and also conserve soil moisture.

(ii) **Hand Weeding:** It is the oldest method of weed control. It is time consuming, labour intensive and often costlier than herbicidal method of weed control. It effectively controls annual weeds, but not perennial weeds. Perennial weeds growing prostrate, rosette and horizontal are frequently cut off at soil surface by this method leaving plant parts (rhizomes, tubers, bulbs, stolons, etc) in the soil, may rejuvenate/ regenerate after some time. Hand weeding may skip / leave unweeded weeds mimics to crop plants and growing within the crop rows. The availability of manual labourers during the peak period of field operation is a big problem for hand weeding.

(iii) Hoeing: Hoeing is a post-planting intercultural operation, very effective against annual weeds but not for perennial weeds, since it cannot control the underground vegetative parts of perennial weeds. Hoeing is comparatively faster operation requiring less man power compared to hand weeding. It may, however, leave some weeds growing within the crop rows. High cost and availability of labour in these days also discourage its adoption. The important implements for hoeing includes, fork, hand hoe, blade harrow, rotary weeder, wheel hoe, power weeder, etc., as given in Fig. 6.1.

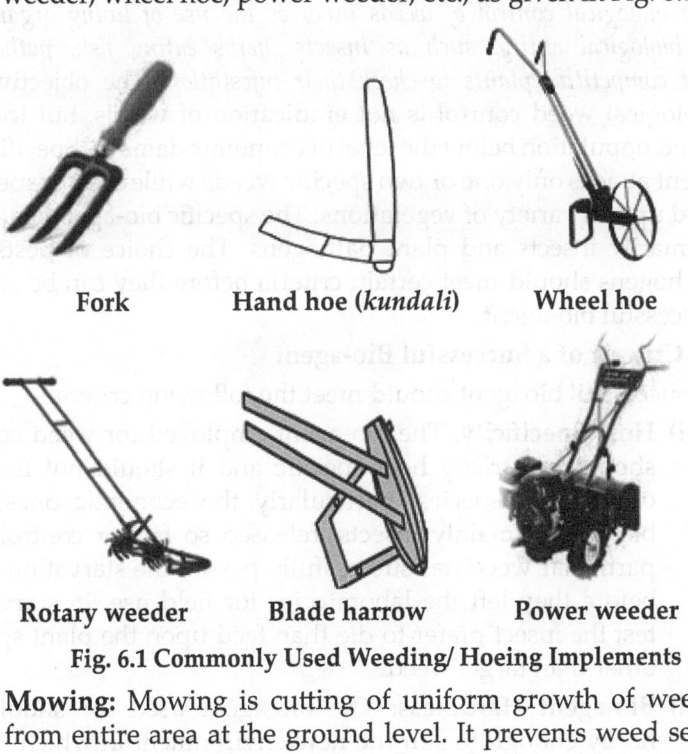

Fig. 6.1 Commonly Used Weeding/ Hoeing Implements

(iv) Mowing: Mowing is cutting of uniform growth of weeds from entire area at the ground level. It prevents weed seed production and dissemination through concurrent control of existing weeds or wild vegetation. Mowing is usually practiced in non-crop areas, lawns, and gardens, but it can also prove useful in removing weeds from rows of certain established crops. Common mowing tools used are sickle, lawn-mower and reciprocating type rotary-bar mower.

(v) Mulching: The weeds are covered artificially with different types of mulches viz., straw, stover, leaves, saw dust and dry grasses, etc. These mulches restricts the aeration and light penetration to weeds and thus, they kill. This method is followed in row crops. Recently use of black plastic.

I also becoming popular as they are very effective for the purpose.

6.2.3 Biological Methods

The biological control of weeds involves the use of living organisms, or biological entity, such as insects, herbs eating fish, pathogens, and competitive plants to check their infestation. The objectives of biological weed control is not eradication of weeds, but to keep weed population below the level of economic damage. Specific bio-agent attacks only one or two specific weeds while the nonspecifics feed upon a variety of vegetations. The specific bio-agents include, primarily insects and plant pathogens. The choice of pests and pathogens should meet certain criteria before they can be said as successful bio-agent.

(a) Criteria of a Successful Bio-agent

A successful bioagent should meet the following criteria–

(i) Host Specificity: The bio-agent employed for weed control should be strictly host specific and it should not feed on other plant species, particularly the economic ones. The bio-agents, mainly insects, released so far for control of a particular weed has successfully passed the starvation tests' before they left the laboratories for field use. In starvation test the insect prefer to die than feed upon the plant species other than target weed.

(ii) Bio-agent Hardiness: The bio-agent used for should be hardy enough to suit the new environment in which it was released. Further the bio-agent should also have the capacity to survive when there is shortage of food for short and long periods.

(iii) **Feeding Habit:** The destructive capacity of a bio-agent on weed should be very high so that its small population could destroy large weed population. Insects feeding on flowers and seeds have been found more host specific and efficient in control of weeds.

(iv) **Ease of Multiplication:** High rate and ease of natural reproduction of bio-agent is important in insects, pathogens, snails, and competitive plants, but very much unwanted in case of grass carp as their excessive population may compete with cultivated fish.

(v) Effective on several taxa of the weed in question.

(b) Kinds of Classical Bio-agents

The bio-agents successful for controlling weeds are as follows:

(i) **Insects:** This is the most important bio-agent frequently used for biological control of weeds. Insects selected as bio-control agents of weeds are largely host specific i.e. one insect species is employed to destroy only one weed species.

(ii) **Plant Pathogens**: Many fungi are observed to attack specific weed species, but examples of successful, practical weed control with fungi are few.

(iii) **Carp Fish:** Certain fresh water carp fish consume large quantities of aquatic weeds and they are effective for making water bodies free from massive growth of noxious, aquatic vegetation.

(iv) **Competitive Plants**: Certain plants may be highly competitive in nature; capable of suppressing some specific weeds without causing any meaningful hinderance to our interests. For example marigold (*Tagetes* spp.) plants have shown potential of displacing *Parthenium hysterophorus.*

(c) Examples of Successful Biological Control of Weeds in different countries are given in table 6.1 as under:

Table 6.1. Biological Control of Weeds by Different Bio-Agents.

Name of the weed	Promising bio-agent	Reporting country	Kinds of bio-agent
Lantana (Lantana camara L.)	*Crocidosema lantana (Busck)*	Hawaii	A moth
Pricklypear (*Opuntia* spp.)	*Cactoblastis cactorum* (Berg)	Australia	An insect
Alligator weed (*Alternanthera philoxeroides Mart.*) Griseb	*Agasides hygrophylla*	Florida (USA)	A flea beetle
Purple nutsedge (*Cyperus rotundus L.*)	*Bactra verutana* (Zeller)	India, USA & Pakistan	A shoot boring moth
Water hyacinth (*Eichhornia crassipes* Mart.) Solms	*Rhizoctonia solani* (Kuehn)	USA	A fungus
Orobanche (*Orobanche cernua Loefl)*	*Sclerotinia spp.*	USA	A fungus
Puncture vine, (*Tribulus terrestris L.*)	*Microlarinus lareynii, M. lypriformis*	USA	A seed pod weevil
Cocklebur (*Xanthium strumarium L.*)	*Nupserha vexator*	–	A beetle
Carrot grass (*Parthenium hysterophorus L.*)	*Zygogramma bicolorata* Pallister	–	*Mexican beetle*
Hydrilla (*Hydrilla verticillata L.F.*) Royle	*Hydrellia pakistanae*	USA	Shoot fly

(d) Advantages and Disadvantages of Biological Weed Control

Advantages:

(i) Biological control is eco-friendly since it does not lead to environmental pollution.

(ii) It is economical in long run, although initial cost is high.

(iii) It is self-perpetuating/self sustaining except bio-herbicide, as bio-herbicide approach is repetitive in nature.

(iv) It sustains bio-diversity as 100 per cent weed control is hardly achieved by use of biological agents.

(v) It is very effective to control target weed, even in dense forests inaccessible to man.

Disadvantages:

(i) Biological method incur high initial cost.

(ii) Possibility of damaging economic crop in case the screening and identification of bio-agents is not proper.

(iii) Very limited use / adoption in field crops.

(iv) Initially very slow in action.

(v) Eradication of weed is not possible by this method.

(vi) Bio-agents very effective to control weeds in a particular country/environment may fail to work in another environment due to adverse weather conditions.

(vii) The active duration (span) of bio-agent in many of the cases is small narrow.

(e) Factors Affecting the Success of Biological Weed Control

(i) Choice of Enemy Agent: The major factors in successful biological control are introduction of a good enemy agent.

(ii) Aggressiveness of Weed: Competitive ability or aggressiveness is affected by favourable climatic and soil conditions, reduced competition from other plant species, and absence of enemies that attack it. Resistance of the host weed to an insect attack is largely determined by its inherent ability to withstand or recover from tissue destruction by insects and pathogens, ability to multiply faster than the attacking agent even under the agent's heaviest attack and greater ability to disperse than that of the enemy.

(iii) Climatic Conditions: Climate affects the adaptability of the insect and competitive ability of the weed host. Ability of an insect to survive extreme conditions and to rebound in numbers before the weed can recover its vigour is an important asset in an enemy agent.

(iv) Competition of Plant Species: The effect of other plant species in the environment on the enemy weakened weed species determine the degree of biological control attained and also the nature of replacement of vegetation.

6.2.3.1 *Use of Bio-herbicides*

Bio-herbicides, in variance with bio-agents, are herbicidal chemicals released by some living organisms. At present these organisms comprise largely certain microbes (fungi/bacteria) because of which bio- herbicides are also termed as myco-herbicides. Bio-herbicides are sprayed in every season on the target weed in crop field. The bio-agent generally remains active only on concurrent weed population. Then they wither away without any cyclic perpetuation unlike what happens in classical bio-control. However, in some cases, the pathogen may remain active for 3 – 4 years e.g. soil-borne pathogen *Phytophthora palmivora* for the control of *Morrenia odorata* in citrus crop.

(a) **Characteristics of Bio-herbicides**

(i) Bio herbicides are living entities / inoculums of plant pathogens mostly fungi.

(ii) They are capable of in-vitro culturing in artificial media and mass production.

(iii) They are applied directly to target weeds.

(iv) They are commercially formulated and applied like herbicides in the field.

(b) **Mode of Damage of Bio-herbicides:** Bio-herbicide spread by rains, wind, insects, contact with host parts, direct vegetative growth and explosive seed discharge. Infection results directly from host susceptible surface, from natural openings or from injured parts. They destroy / degrade cell constituents / organelles enzymatically. They produce toxins, disturb hormonal system, obstruct translocation of foods and minerals and render malfunctioning of physiological processes. This results in death of weed plants or weed plants fall weak to compete with crop plants.

(c) **Commercial Myco-herbicides in Use in the World** are listed in table 6.2 as under:

Table 6.2. Commercial Myco-herbicides

S.No.	Product	Content	Weed controlled
1	De-vine	Liquid suspension of fungal spores of *Phytophthora palmivora.* It causes root rot in the weed.	*Morrenia odorata* (strangler vine) in citrus orchards
2	Collego	Wettable powder contains fungal spores of *Colletotrichum gloesporiodes* sub spp. *ecschynomone*	*Aeschynomone* spp. (Joint-vetch) in rice fields by causing stem and leaf blight in the weed
3	Biomal	*Collectotrichum gloesporiodes* sub spp. malue	*Malva passilia* (round leaved mallow) in wheat and small grain crops
4	Velgo	*Collectotrichum coccoides*	*Abutilon theophrasti* in cotton
5	Bipolaris	A suspension of fungal spores of *Bipolaris sorghicola*	*Sorghum halepense* (Johnson grass)
6	Lubao	*Colletotrichum gloesporiodes* spp. *cuscutae*	*Cuscuta spp.* (dodder)

(d) Advantages and Disadvantages of Bio-herbicides

Advantages

(i) A high degree of specificity of bio-herbicides to target weeds.
(ii) No health hazards to human beings and animals.
(iii) No environmental pollution.
(iv) Effective control of perennial, parasitic and invasive weeds in non-cropped areas.
(v) No development of herbicide resistance in weeds.

Disadvantages

(i) A slow process of controlling / killing of weeds.
(ii) A large number of bio-herbicides are required to control a mixed population of weeds in a crop.
(iii) Environment and ecology greatly affect their stability and effectiveness.

(iv) The shelflife (storage life) is quite less compared to herbicides.

(v) High production / multiplication cost involved.

(vi) Bio-herbicides need registration with Environmental Protection Agency (EPA), which is a long time consuming process.

(vii) Early weed competition may damage the crop as bio-herbicide is applied only when weed attains certain level of growth / canopy coverage.

6.2.4 Integrated Weed Management (IWM)

It has been realized that for sustainable agriculture we must develop and practice the concept of integrated weed management as against the conventional 'weed control' approach. While weed control aims at only putting down the weeds already present by some kind of physical or chemical energy, integrated weed management is a system approach whereby a whole land use planning is done in advance to minimize the very invasion of weeds in aggressive forms and give crop plants a strong, competitive advantage over the weeds.

IWM although necessarily a combination of direct and indirect methods of weed control, it, however, does not mean that all methods have to be considered together, but it necessarily should be an integration of more than one method based on their suitability, compatability and feasibility under a specific crop-weed situation. Integrated weed management, therefore, varies from place to place, from crop to crop, from season to season and even based on soil properties. In organic farming use of herbicides or any chemical means of weed control is not included in integrated weed management. While integrating different methods of weed control following points may be taken care of-

(i) The methods selected for weed management should be mutually compatible to each other and also match the diversity of weed species present in the crop field or a situation.

(ii) They should be favourable to crop growth, but unfavorable to weeds.

(iii) It should give effective weed control for longer duration under certain situation.

(a) Major Principles of IWM

(i) Manipulation of crop habitat by utilizing some biological differences between the crops and weeds, with the objective to place the former in a competitive advantage over the later.

(ii) Measures should be directed to break the survival mechanism of the weed in the soil.

(iii) Good crop husbandary practices should be incorporated to avoid establishment of perennial and parasitic weeds.

(iv) Any individual practice of weed management should be eco-friendly as far as possible.

(b) Components of Integrated Weed Management (IWM): The components of integrated weed management in organic farming includes preventive measures, good crop husbandry or cultural methods, mechanical and manual methods, and biological methods. These are the broader components of IWM, however, we need to formulate a workable integration of these methods considering the climate, soil, crop, nature of weed infestation, critical period of weed crop competition, availability of inputs and socio-economic conditions of farmers. Several options / methods such as competitive and well suited high yielding varieties to resist weed competition, summer ploughing, pre-sowing seed bed tillage, stale seed bed preparation, manual weeding (hand tools, weeder), mechanical weeding (animal / power drawn weeders), changing in cropping system / better crop management techniques (seeding method, optimum plant population, precise use of water, use of bio-agents, mulching, soil solarization, etc. may be tried towards possible integration. Such integration will greatly improve weed management in crops and cropping systems in organic farming.

(c) Advantages and Disadvantages of IWM

Advantages/ usefulness

It is more effective approach in a long run as different methods of weed control will take care of weeds in totality vis-a-vis checks seed production of weeds.

(i) It is environment or eco-friendly.

(ii) It may generate higher net income / return in long run particularly under high cropping intensity.

(iii) It is very useful under high cropping intensity.

(iv) It is developed for whole farm and thereby prevents possible dispersion of weeds from adjoining uncultivated areas.

(v) It may reduce or eliminate chance of occurrence of weed flora shift, preponderance of perennial weeds, etc. in an agro-ecosystem.

Disadvantages:

(i) It is highly site specific and crop or cropping system specific.

(ii) It is more of a concept and as such not a method of weed control.

(iii) Integration of different methods of weed control may incur higher cost.

(iv) A IWM model designed for certain situations may not be operative constantly or repetitive over the years.

Chapter 7

Organic Crop Production

Use of inputs in organic farming is altogether different from the conventional farming as the use of chemicals for plant nutrition as well as insects pests, diseases and weed control is totally prohibited. In organic farming only organic origin inputs are used with the objective to check the environmental pollution and health hazards. In this chapter crop production technologies will be discussed which are pertinent to organic farming in light of the procedures and standards fixed for certification of organic produce. The organic crop production technologies are described as follows.

7.1 DURATION OF CONVERSION PERIOD

The establishment of an organic management system and building of soil fertility requires an interim period, the conversion period. The **conversion period** may not always be of sufficient duration to improve soil fertility and reestablish the balance of the ecosystem but it is the period in which all the actions required to reach these goals are started.

Plant products produced can be certified organic when the national standards requirements have been met during a conversion period of at least two years before sowing or in the case of perennial crops other than grassland, at least three years (thirty-six months) before the first harvest of products. The accredited inspection and certification agency may decide in certain cases (such as idle use for two years or more) to extend or reduce the conversion period in the light of previous status of the land but the period must equal or exceed twelve months.

The conversion period can be extended by the certification programme depending on, e.g., past use of the land and environmental conditions.

The certification programme may allow plant products to be sold as "produce of organic agriculture in process of conversion" or a similar description during the conversion period of the farm.

For the calculation of inputs for feeding, the feed produced on the farm unit during the first year of organic management, may be classified as organic. This refers only to feed for animals which are themselves being reared within the farm unit and such feed may not be sold or otherwise marketed as organic. Feed produced on the farms in accordance with the national standards is to be preferred over conventionally grown / brought-in feeds.

7.2 CHOICE OF CROPS AND VARIETIES

Selection of crop depends mainly on prevailing farm conditions, adaptability to the prevailing conditions, marketability and profitability, resistance to pests and diseases, prevailing farming system, etc. While selecting the crops and variety in organic farming following points may be taken care:

(i) All seeds and plant material shall be certified organic. Species and varieties cultivated shall be adapted to the soil and climatic conditions and be resistant/ tolerant to pests and diseases. In the choice of varieties, genetic diversity shall be taken into consideration.

(ii) Preferably certified organic seed and plant materials shall be used. If certified organic seed and plant materials are not available, chemically untreated conventional seed and plant material shall be used.

(iii) The use of genetically engineered seeds, transgenic plants or plant material is prohibited.

7.3 DIVERSITY IN CROP PRODUCTION

(i) The basis for crop production in organic farming shall take into consideration the physical condition and fertility of the soil and the surrounding ecosystem, with a view to minimize the nutrient losses. Diversity in crop production is achieved by a combination of a versatile crop rotation with legumes and an appropriate coverage of soil during the year of production which diverse plant species.

(ii) Where appropriate, the organic farms shall be required to maintain sufficient diversity in a manner that takes into account pressure from insects, weeds, diseases and other pests, while maintaining or increasing soil organic matter, fertility, microbial activity and general soil health. For non perennial crops, this is normal, but not exclusively achieved by means of crop rotation preferably by leguminous crops.

7.4 NUTRIENT MANAGEMENT

For nutrient management in organic ecosystem following points may be taken into consideration.

(i) Sufficient quantities of biodegradable materials of microbial, plant or animal origin produced on organic farms shall form the basis of the nutrient management programme to increase or at least maintain its fertility and the biological activities within it.

(ii) Fertilization management should minimize nutrient losses. Accumulation of heavy metals and other pollutants shall be prevented.

(iii) Non synthetic mineral fertilisers and brought-in bio-fertilisers (biological origin) shall be regarded as supplementary and not as a replacement for nutrient recycling.

(iv) The certification programme shall set limitations to the total amount of bio-degradable material of microbial, plant or animal origin brought onto the farm unit, taking into account local conditions and the specific nature of the crops.

(v) The certification programme shall set procedures which prevent animal runs from becoming over manuring where there is a risk of pollution.

(vi) Mineral fertilizers shall only be used in a supplementary role to carbon based materials. Only those organic or mineral fertilizers that are brought in to the farm (including potting compost) shall be used when, the circumstances demand in accordance with the standards.

(vii) Permission for use shall only be given when other fertility management practices have been optimized.

(viii) Manures containing human excreta (faeces and urine) shall not be used.

(ix) Mineral fertilisers shall be used in their natural composition and shall not be rendered more soluble by chemical treatment. The certification programme may grant exceptions. These exceptions shall not include mineral fertilizers containing nitrogen.

(x) The certification programme shall lay down restrictions for the use of inputs such as mineral potassium, magnesium fertilisers, trace elements, manures and fertilisers with a relatively high heavy metal content and/or other unwanted substances, e.g. basic slag, rock phosphate and sewage sludge, chilean nitrate and all synthetic nitrogenous fertilisers including urea are prohibited/restricted for use.

7.5 INSECT PESTS, DISEASE AND WEED MANAGEMENT

To manage insect pests, diseases and weeds in organic crop production following points may be considered.

(i) Organic farming systems shall be carried out in a way which ensures that losses from pests, diseases and weeds are minimized. Emphasis is given on the use of a balanced fertilizing programme, use of crops and varieties well-adapted to the environment, fertile soils of high biological activity, suitable crop rotations, intercropping, green manures, etc.

(ii) Weeds, insect pests and diseases shall be controlled through a number of preventive cultural techniques e.g. use of weed free seed, use of disease free seed, suitable crop rotations, green manuring, stale seedbed preparations, mulching, mechanical/physical control and the disturbance of pest development cycles, etc. Accredited certification programmes shall ensure that measures are in place to prevent transmission of pests, parasites and infectious agents.

(iii) Pest management shall be regulated by understanding and disrupting the ecological needs of the pests. The natural enemies of pests and diseases shall be protected and encouraged through proper habitat management of hedges, nesting sites, etc. An ecological equilibrium shall be created to bring about a balance in the pest predator cycle.

(iv) Products used for insect pests, diseases and weed management, prepared at the farm from local plants, animals and micro-organisms, shall be allowed. If the eco-system or the quality of organic products might be jeopardized, the certification programme shall judge if the product is acceptable as per the procedure given to evaluate additional inputs to organic agriculture.

(v) Thermic weed control and physical methods for pest, disease and weed management shall be permitted.

(vi) Thermic sterilization of soils to combat pests and diseases shall be restricted to circumstances where a proper rotation or renewal of soil cannot take place. The certification programme on a case-by-case basis may only give permission.

(vii) All equipments from conventional farming systems shall be properly cleaned and free from weed seeds and diseases and insect inoculum before being used on organically managed areas.

(viii) The use of synthetic insecticides, herbicides, fungicides, growth regulators, synthetic dyes, and other pesticides are prohibited. Permitted products for insect pests and disease control are to be used only.

(ix) The use of genetically engineered organisms or products is prohibited.

7.6 CONTAMINATION CONTROL

To minimize all relevant contamination from outside and within the farm following measures shall be taken –

(i) Buffer zones shall be maintained to prevent contamination from conventional farms. The buffer zone should be sufficient in size to prevent the possibility of unintended contact of prohibited substances applied to adjacent conventional land areas/farms.

(ii) In case of reasonable suspicion of contamination, the certification programme shall make sure that an analysis of the relevant products and possible sources of pollution (soil and water) shall take place to determine the level of contamination.

(iii) Polyethylene and polypropylene or other polycarbonates coverings such as plastic mulches, fleeces, insect net and silage wrapping, only are allowed. These shall be removed from the soil after use and shall not be burnt on the farmland. The use of polychloride based products is prohibited.

7.7 SOIL AND WATER CONSERVATION

Soil and water are most important natural resources for crop production, therefore this shall be handled in a sustainable manner. For this following points may be considered.

(i) Relevant measures shall be taken to prevent erosion, salination of soil, excessive and improper use of water and the pollution of ground and surface water.

(ii) Clearing of land through the means of burning organic matter, e.g. slash-and-burn, straw burning shall be restricted to the minimum. The clearing of primary forest is prohibited.

(iii) The certification programme shall require to check appropriate stocking rates which does not lead to land degradation and pollution of ground and surface water.

Chapter

8

Certification Process and Standards of Organic Farming in India

Organic farming has a major difference from other approaches to sustainable agriculture, is the existence of production standards. Certification procedures and in many countries a legislative basis giving a clear dividing line between organic and other farming systems, primarily for marketing purposes. For products to be labelled as organic they must be currently be certified by third party organization as having been produced according to specific certification standards, shaped over time and around the world by hundreds of certifying, non-government organizations and more recently by national governments.

8.1 ORGANIC CERTIFICATION

Organic certification is a certification process for producers of organic food and other organic agricultural products. In general, any business directly involved in food production can be certified, including seed suppliers, farmers, food processors, retailers and restaurants. Requirements vary from country to country, and generally involve a set of production standards for growing, storage, processing, packaging and shipping that include:

- Avoidance of synthetic chemical inputs (e.g. fertilizer, pesticides, antibiotics, food additives, etc.) and genetically modified organisms;
- Use of farmland that has been free from chemicals for a number of years (often three or more);

- Keeping detailed written production and sales records (audit trail);
- Maintaining strict physical separation of organic products from non-certified products;
- Undergoing periodic on-site inspections.

In some countries, certification is overseen by the government, and commercial use of the term organic is legally restricted. Certified organic producers are also subject to the same agricultural, food safety and other government regulations that apply to non-certified producers.

8.2 PURPOSE OF CERTIFICATION

Organic certification addresses a growing worldwide demand for organic food. It is intended to assure quality and prevent fraud. For organic producers, certification identifies suppliers of products approved for use in certified operations. For consumers, "certified organic" serves as a product assurance, similar to "low fat", "100% whole wheat", or "no artificial preservatives".

Certification is essentially aimed at regulating and facilitating the sale of organic products to consumers. Individual certification bodies have their own service marks, which can act as branding to consumers. A certifier may promote the high consumer recognition value of its logo as a marketing advantage to producers. Most certification bodies operate organic standards that meet the national government's minimum requirements.

8.3 THE CERTIFICATION PROCESS

8.3.1 Third Party Certification

In third party certification, the farm or the processing of the agriculture produce is certified in accordance with national or international organic standards by an accredited organic certification agency. In order to certify a farm, the farmer is typically required to engage in a number of new activities, in addition to normal farming operations as under–

- **Study** the organic standards, which cover in specific detail what is and is not allowed for every aspect of farming, including storage, transport and sale.
- **Compliance** - Farm facilities and production methods must comply with the standards, which may involve modifying facilities, sourcing and changing suppliers, etc.
- **Documentation** - Extensive paperwork is required, detailing farm history and current set-up, and usually including results of soil and water tests.
- **Planning** - A written annual production plan must be submitted, detailing everything from seed to sale: seed sources, field and crop locations, fertilization and pest control activities, harvest methods, storage locations, etc.
- **Inspection** - Annual on-farm inspections are required, with a physical tour, examination of records, and an oral interview.
- **Fee** – A fee is to be paid by the grower to the certification body for annual survellence and for facilitating a mark which is acceptable in the market as symbol of quality.
- **Record-keeping** - Written, day-to-day farming and marketing records covering all activities must be available for inspection at any time. In addition, short-notice or surprise inspections can be made, and specific tests (e.g. soil, water, plant tissue) may be requested.

  For first-time farm certification, the soil must meet basic requirements of being free from use of prohibited substances (synthetic chemicals, etc.) for a number of years. A conventional farm must adhere to organic standards for this period, often, three years. This is known as being in transition. Transitional crops are not considered fully organic. A farm already growing without chemicals may be certified without this delay

Certification for operations other than farms is similar. The focus is on ingredients and other inputs, and processing and handling conditions. A transport company would be required to detail the use and maintenance of its vehicles, storage facilities, containers, and so forth. A restaurant would have its premises inspected and its suppliers verified as certified organic.

Certification Procedure in Brief

- Application is made to the certification agency in the prescribed format with necessary farm and process details
- Screening of application by certification agency and if necessary further details/certification sought.
- Cost estimate comprising certification charge, inspection charge, travel cost, reporting cost, laboratory charges, etc. is sent for acceptance of cost by the grower/ producer.
- Signing of agreement between grower/ producer and certification agency.
- Certification agency seeks cropping/production/cultivation/ processing plan and supply a copy of the standards to the grower/producer to follow.
- Certification agency raises an invoice and asks the producer to release 50 per cent of the certification cost in advance.
- Grower/ producer pays the fee.
- Inspection schedule is worked out.
- Inspection is carried out at one or more than one occasion.
- If required unannounced inspection can also be done. In case of doubt the inspection team can also draw plant/soil/ raw material/input/product sample for laboratory analysis.
- Inspection report (s) submitted to the certification committee
- Certification agency asks for final payment
- Certification is granted

Grower/producer releases the stock for sale with certification mark (India organic logo).

8.3.2 Grower Group Certification System

The Organic Agriculture movement will not be successful until it reaches the small holder farmers. Previously, Organic Standards and Certification procedures were designed considering only individual farmers, the potential for smallholder groups certification being realized latter. Group Certification has now been introduced and is most popular and acceptable amongst the Indian farming community. Small holder group certification brings the following benefits:

(i) It reduces certification cost per hectare per year compared to individual farmers certification.

(ii) Small holder groups gathering farmers with common interests make it possible and easy for government to implement development schemes.

(iii) It helps the farmers in collective bargaining with buyers.

(iv) It saves transportation cost incurred by group members in transporting their produce to market.

(v) National and international buyers get bulk volume of particular commodity.

(vi) It ensures a regular supply of consistent quality to buyers and processors.

(vii) It provides well opportunity for input providers to market among group members.

(viii) It helps in knowledge /information dissemination among large number of farmers.

(ix) It contributes to maintenance of social harmony as groups may contain farmers of distinct caste and religion.

(x) The growers organized into groups become an easy market for certification agencies.

(xi) Members can benefits from group training in Organic Agriculture, Biodynamic agriculture, Organic Certification, Organic produce quality, demonstrations, tours , etc.,

(xii) All groups members can receive advice on seeds, fertility, pests and diseases, post harvest related issues and marketing

8.3.2.1 Who can be considered as a group?

(i) Only small farmers can be members of the group covered by group certification. Large farms can also belong to the group but have to be inspected annually by the external inspection body. Processors and exporters can be part of the structure of the group, but have to be inspected by the external inspection body.

(ii) The farmers of the group apply similar production system and the farms should be in geographical proximity.

(iii) A group may be organised on itself i.e., as a cooperative, or as a structural group of producers affiliated to a processor or an exporter.

(iv) The group must be established formally, based a written agreement with its members. It shall have control management, established decision procedures and leagel capacity.

(v) When intended for export, the marketing of the products must be carried as a group.

8.3.2.2 Principles

- A substantial part of the inspection work is carried out by internal inspections in the frame work of the internal control system set up by the group.
- The external inspection body verifies and evaluates the effectiveness of the internal control system and certifies the group as a whole.

8.3.2.3 Internal Quality System (IQS)

Group certification is based on the concept of an Internal Quality System comprising the followings:-

- Implementation of the internal control system
- Internal standards
- Risk assessment

An external inspection and certification body should be identified for conducting annual inspection of the individual group / unit. The external inspection agency shall evaluate by checking the IQS documentation, staff qualifications and re-inspecting some farms.

(i) **Developing Internal Quality System:** The following are the minimum requirements for setting up an IQS for grower groups:

- Development of Internal Control System (ICS)
- Identification of producer groups
- Creation of awareness about group certification
- Identification of qualified personnel for maintaining the ICS
- Give necessary training in production and IQS development
- Preparation of IQS manual containing policies and procedures
- Implementation of the policies and procedures
- Review and improvement of the IQS document for maintaining a harmonized IQS.

(ii) The System is Operated Through Following Workers

- Internal quality system manager (IQS Manager)
- Internal inspectors
- Approved manager / committee
- Field officers
- Purchase officers
- Warehouse manager
- Processing manager (in case of processing unit)

(iii) Internal Standards: The internal standards shall be prepared in local language by the IQS manager for the region of operations under the framework of NPOP standards. If the farmers are illiterate, the internal standards shall be contain illustrations in the text for better understanding.

(iv) Conflict Interest: The IQS personnel shall not have any conflict of interest that might hinder the work. All possible conflicts shall be declared a written statement. In such cases, the IQS shall ensure that alternative solutions found.

(v) Scope of Certification: The certification shall be granted to the group with reference to the regulations / standards adopted by the group.

(vi) Procedure for Implementation of IQS–

- Registration of members – All members of the group will be formally (legally) registered under a single entity.

- Provision of documents to the members of the grower group – Each member of the grower group will be supplied with docket in local languages, which will contain – copy of IQS manual, internal standards document, NPOP document, definition of the production unit, farm entrance form, field records, written contract, annual farm inspection checklist and information on training programmes and provision of advisory services.

(vii) **Internal Inspections:** At least two inspections of the group (one in growing season of each crop) shall be carried out by the internal inspector and will be documented. The inspection will be carried out in presence of the member or his representative and must include a visit of the whole farm, storage of inputs, harvested products, post harvest handling and animal husbandry. In case of non-compliance, the results will be reported to the IQS manager and all measures should be taken according to the internal sanction procedures.

(viii) **External Inspections:** The external inspection and Certification agency will re-inspect some of the farms for the evaluation of the grower group for efficient internal control system for compliance with the NPOP standards. The sampling plan for inspection shall be based on the inspector's perception of risk.

(ix) **Yield Estimates:** Yields will be estimated for each crop for individual farmer in the group. This activity should be carried out especially during harvesting and should be counter checked with the estimates during buying.

8.3.3 Participatory Certification

Participatory Guarantee Systems (PGS) are locally focused quality assurance systems. They certify producers based on active participation of stakeholders and are built on a foundation of trust, social networks and knowledge exchange (Definition by IFOAM – organics international). Over the years, PGS have proved to be an affordable alternative to third – party certification particularly appropriate for small -scale farmers who cannot cover the costs of other forms of certification. PGS open a way for them to access local markets for organically produced food. PGS have spread impressively in the past few years. However, this has not come without challenges. Respondents to the Global PGS Survey 2015 issued by IFOAM – Organics international have pointed the following as main issues they have been confronted with while developing PGS:

(i) As far as the logistics are concerned, farmers are generally distant from each other and from the market place, organic inputs are not always easily available and accessible, and infrastructures (e.g., communication tools, transportation facilities) are not developed enough to ease interactions within the PGS and towards the market.

(ii) Lack of consumer awareness and lower visibility of PGS result in the limitation of market opportunities for the involved producers. This is also a consequence of difficulties faced in diversifying production and keeping it constant over the year. A further challenge exists in finding the human and financial resources to conduct field visits.

(iii) Finally, governments may have an adverse impact on the development of PGS when their organic regulations do not recognise PGS or when they over – bureaucratise the conditions for their recognition.

Participatory Guarantee Systems (PGS) represent an alternative to third party certification. Participatory certification is especially adapted to local markets and short supply chains. They can also complement third party certification with a private label that brings additional guarantees and transparency. Participatory guarantee system (PGS) enable the direct participation of producers, consumers and other stakeholders in;

- The choice and definition of the standards
- The development and implementation of certification procedures
- The certification decisions

Participatory Guarantee Systems are also referred to as "participatory certification".

8.3.4 Alternative Certification

The word organic is central to the certification (and organic food marketing) process, and this is also questioned by some. Where organic laws exist, producers cannot use the term legally without certification. To bypass this legal requirement for certification, various alternative certification approaches, using currently undefined terms like "authentic" and "natural", are emerging. In the US, motivated by the cost and legal requirements of certification (as of October, 2002), the private farmer – to farmer association, Certified naturally grown offers a " non-profit alternative eco-labelling program for small farms that grow using USDA Organic Methods but are not a part of the USDA Certified Organic Program.

8.4 CERTIFICATION AROUND THE WORLD

In some countries, organic standards are formulated and overseen by the government. The United States, the European Union, Canada and Japan have comprehensive organic legislation, and the term "organic" may be used only by certified producers. Being able to put the word "organic" on a food product is a valuable marketing advantage in today's consumer market, but does not guarantee, the product is legitimately organic. Certification is intended to protect consumers from misuse of the term, and make buying organics easy. However, the organic labelling made possible by certification itself usually requires explanation. In countries without organic laws, government guidelines may or may not exist, while certification is handled by non-profit organizations and private companies.

Internationally, equivalency negotiations are underway, and some agreements are already in place, to harmonize certification between countries, facilitating international trade. There are also international certification bodies, including members of the International Federation of Organic Agriculture Movements (IFOAM) working on harmonization efforts. Where formal agreements do not exist between countries, organic product for export is often certified by agencies from the importing countries, who may establish permanent foreign offices for this purpose.

In the US, the National Organic Program (NOP), was enacted as federal legislation in October 2002. It restricts the use of the term "organic" to certified organic producers (excepting growers selling under $5,000 a year, who must still comply and submit to a records audit if requested, but do not have to formally apply). Certification is handled by state, non-profit and private agencies that have been approved by the US Department of Agriculture (USDA).

In Japan, the Japanese Agricultural Standard (JAS) was fully implemented as law in April 2001. This was revised in November 2005 and all JAS certifiers were required to be re-accredited by the Ministry of Agriculture.

In India, APEDA regulates the certification of organic products as per National Standards for Organic Production. The NPOP standards for production and accreditation system have been recognized by European Commission and Switzerland as equivalent to their country standards. Similarly, USDA has recognized NPOP conformity assessment procedures of accreditation as equivalent to that of US.

With these recognitions, Indian organic products duly certified by the accredited certification bodies of India are accepted by the importing countries. In March 2000, the Ministry of Commerce launched NPOP (National Programme for Organic Production) design to establish national standards for organic products which could then be sold under the logo India Organic. For proper implementation of NPOP, NAPP (National Accreditation Policy and Programme) has been formulated, with Accreditation Regulations announced in May 2001. These make it mandatory that all certification bodies whether internal or foreign operating in the country must be accredited by an Accreditation Agency. The regulations make provision for export, import and local trade of organic products. However, currently only the exports of organic products come under government regulations. Thus an agricultural product can only be exported as an organic product if it is certified by a certification body duly accredited by APEDA. Organic crop production, organic animal production, organic processing operations, forestry and wild products are the categories of products covered under accreditation.

In China, the China Green Food Development Center awards two Standards: A and AA; while the former standard does permit some use of synthetic agricultural chemicals, the latter is more stringent.

8.5 ORGANIC CERTIFICATION SYSTEM IN INDIA

In India, there are two accreditation systems for authorizing Certification and Inspection agencies for organic certification. National Programme on Organic Production (NPOP) promoted by Ministry of Commerce is the core programme which governs and defines the standards and implementing procedures. National Accreditation Body (NAB) is the apex decision making body. Certification and inspection agencies accredited by NAB are authorized to undertake certification process. The NPOP notified under FTDR Act and controlled by Agricultural Processed Food Export Development Authority (APEDA) looks after the requirement while NPOP notified under APGMC Act land controlled by Agriculture Marketing Advisor, Directorate of Marketing and Inspection looks after domestic certification process.

8.5.1 National Programme on Organic Production (NPOP)

To provide a focused and well directed development of organic agriculture and quality products, Ministry of Commerce and Industry, Government of India, launched a National Program on Organic Production in the year 2000, which was formally notified in October 2001 under the Foreign Trade & Development Act (FTDR Act). This document provides information on standards for organic production, systems criteria, and procedures for accreditation of Inspection and Certification bodies, the national organic logo and the regulations governing its use. The standards and procedures have been formulated in harmony with international standards such as those of Codex and IFOAM.

8.5.1.1 Scope and Operational Structure of NPOP:

The National Programme for Organic Production proposes to provide an institutional mechanism for the implementation of National Standards for Organic production, through a National Accreditation Policy and Programme. The aims of the National Programme for organic production, inter alia, includes the following:

- To provide the means of evaluation of certification programmes for organic agriculture and products as per the approved criteria.
- To accredit certification programmes.
- To facilitate certification of organic products in conformity to the National Standards for Organic Products.
- To encourage the development of organic farming and organic processing.

(i) Scope: The National Programme for Organic Production shall, among others, includes:

- Policies for development and certification of organic products
- National standards for organic products and processes
- Accreditation of programmes to be operated by Inspection and Certification Agencies
- Certification of organic products

(ii) Operational Structure: The programme will be developed and implemented by the Government of India through its Ministry of Commerce and Industry as the apex body. The Ministry will constitute a National Steering Committee for National Programme for Organic Production, whose members will be drawn from Ministry of Commerce and Industry, Ministry of Agriculture, Agricultural and Processed Food Products Export Development Authority (APEDA), Coffee Board, Spices Board, Tea Board and other government and private organizations associated with the organic movement. To advise the National Steering Committee on relevant issues pertaining to National Standards and Accreditation, sub-committees will be appointed. The National Steering Committee for National Programme for

Organic Production will formulate a National Accreditation Policy and Programme and draw up National Standards for Organic Products, which will include standards for organic production and processes as well as the regulations for use of the National Organic Certification Mark. National Accreditation Policy and Programme will be administered by the National Accreditation Body, which will define the overall policy objectives for the Accreditation programmes and operations. The National Steering Committee may amend the Accreditation procedures whenever it deems fit. The National Accreditation Policy and Programme is subject to periodic internal review, which will be conducted by the Technical Committee, which will advise the National Steering Committee about the need and content of such amendments in the National Programme for Organic Production.

The operational structure of the National Programme for Organic Production is given in Fig. 8.1.

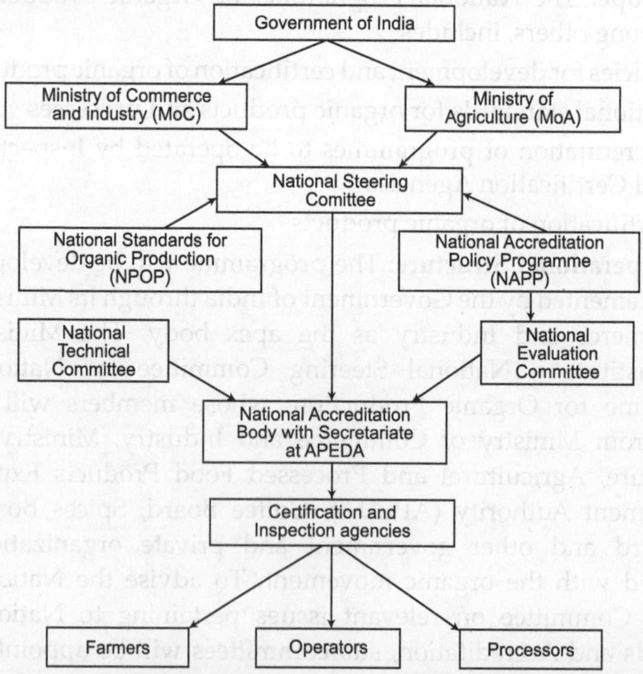

Fig. 8.1 Operational Structure of National Programme for Organic Production (NPOP).

8.6 CERTIFICATION AND PRODUCT LABELLING

Being able to put the word "organic" on a food product is a valuable marketing advantage in today's consumer market. Certification is intended to protect consumers from misuse of the term, and make buying organics easy. However, the organic labelling made possible by certification itself usually requires explanation. In many countries organic legislation defines three levels of organics. Products made entirely with certified organic ingredients and methods can be labelled "100% organic". Products with 95 per cent organic ingredients can use the word "organic". Both may also display organic seal. A third category, containing a minimum of 70 per cent organic ingredients, can be labelled "made with organic ingredients". In addition, products may also display the logo of the certification body that approved them. Products made with less than 70 per cent organic ingredients cannot advertise this information to consumers and can only mention this fact in the product's ingredient statement.

8.6.1 Organic Certification Mark

8.6.1.1 Organic Logo

A trademark – "India Organic" will be granted on the basis of compliance with the National Standards for Organic Production (NSOP). Communicating the genuineness as well as the origin of the product, this trademark is owned by the Government of India. Only such exporters, manufacturers and processors whose products are duly certified by the accredited inspection and certification agencies, will be granted the license to use of the logo which would be governed by a set of regulations.

8.6.1.2 Concept and Colour Specification of Organic Logo

Symbolizing the rhythm of cosmic and earth forces represented by the blue and brown waves of force and energy, 'India Organic' logo celebrates the essence of nature. These forces work in harmony upon the earth's environment and this rhythm is reinforced and supported by the green plant growth. The colours used have a special significance in the logo concept. The cosmic force in blue symbolizes universal purity. Richness of soil, nourished with natural ingredients in organic farming, is symbolized by the earth forces in golden brown. The plant in green uses the colour of nature and natural products untouched by chemicals. The blue background is symbolic of earth's environment that is congenial for life to thrive in and is also free of pollution and harmful chemicals. India Organic etched over the surface authenticates the carrier as "Organic" and also establishes the Indian connection for all the carriers of the mark. Beautifully synthesizing all the elements of our environment, the logo also communicates total adherence to the National Organic Standards.

8.7 NATIONAL STANDARDS FOR ORGANIC PRODUCTION (NSOP)

Standards requirements for crop production, food processing and handling are described in detail in chapter 7 of this book, however, a brief is given here as under–

8.7.1 Conversion Requirements

The time between the start of organic management and cultivation of crops or animal husbandry is known as the conversion period. All standard requirements should be met during conversion period. Full conversion period is not required where organic farming practices are already in use.

8.7.2 Crop Production

8.7.2.1 Choice of Crops and Varieties

All seeds and planting materials should be certified organic. If certified organic seed or planting material is not available then chemically untreated conventional material can be used. Use of genetically engineered seeds, pollen, transgenic plants are not allowed.

8.7.2.2 Duration of Conversion Period

Plant products produced can be certified organic when the national standards requirements have been met during a conversion period. For perennial plants (excluding pastures and meadows) the conversion period is atleast three years before the first harvest of products. Depending upon the past use of the land and ecological situations, the certification agency can extend or reduce the minimum conversion period.

8.7.2.3 Diversity in Crop Production Standards

Where appropriate, the certification programme shall require that sufficient diversity is obtained in time or place in a manner that takes into account pressure from insects, weeds, diseases and other pests, while maintaining or increasing soil organic matter, fertility, microbial activity and general soil health. For non perennial crops, this is normally, but not exclusively, achieved by means of crop rotation.

8.7.2.4 Fertilization Policy

Biodegradable material of plant or animal origin produced on organic farms should form the basis of the fertilization policy. Fertilization management should minimize nutrient losses, avoid accumulation of heavy metals and maintain the soil pH. Emphasis should be given to generate and use own on --farm organic fertilizers. Brought in fertilizers of biological origin should be supplementary and not a replacement. Over manuring should be avoided. Manures containing human excreta should not be used on vegetation for human consumption.

8.7.2.5 Insect Pests, Disease and Weed Management Including Growth Regulators

Weeds, pests and diseases should be controlled preferably by preventive cultural techniques. Botanical pesticides prepared at farm from local plants, animals and microorganisms are allowed. Use of synthetic chemicals such as fungicides, insecticides, herbicides, synthetic growth regulators and dyes are prohibited. Use of genetically engineered organisms or products is prohibited.

8.7.2.6 Contamination Control Standards

In case of reasonable suspicion of contamination, the certification programme shall make sure that an analysis of the relevant products to detect the possible sources of pollution (soil and water), shall take place to determine the level of contamination.

For protected structure coverings, plastic mulches, fleeces, insect netting and silage rapping, only products based on polyethylene and polypropylene or other polycarbonates are allowed. These shall be removed from the soil after use and shall not be burnt on the farmland. The use of polychloride based products are prohibited.

8.7.2.7 Soil and Water Conservation

Soil and water resources should be handled in a sustainable manner to avoid erosion, salination, excessive and improper use of water and the polluting of surface and ground water, cleaning of land by burning of straw, crop residues, etc should be restricted. Clearing of primary forest for agriculture (jhuming or shifting cultivation) is strictly prohibited.

8.7.3 Collection of Non-Cultivated Material of Plant Origin and Honey

Wild harvested products shall only be certified organic, if derived from a stable and sustainable growth environment and the harvesting shall not exceed the sustainable yield of the eco-system and should not threaten the existence of plant or animal species. The collection area should not be exposed to prohibited substance and should be at an appropriate distance from conventional farming, human habitation, and places of pollution and contamination.

8.7.4 Food Processing and Handling

Organic products shall be protected from co-mingling with non organic products and shall be adequately identified through the whole process. Certification programme shall regulate the means and measures to be allowed or recommended for decontamination, clearing or disinfection of all facilities where organic products are kept, handled, processed or stored. Besides storage at ambient temperature the following special condition of storage are permitted.

- Controlled atmosphere
- Cooling,
- Freezing,
- Drying and humidity regulation,
- Ethylene gas is permitted for ripening.

8.7.5 Pest and Disease Control

For pest management and control following measures shall be used in order of priority.

- Preventive methods such as disruption and elimination of habitat and access to facilities.
- Mechanical, physical and biological methods.
- Permitted pesticidal substances as per the standards and other substances used in traps.

Irradiation is prohibited. Direct or indirect contact between organic products and prohibited substances (such as pesticides) should not be there.

8.7.6 Packaging

Materials used for packaging shall be eco-friendly. Unnecessary packaging material should be avoided. Recycling and reusable systems should be used. Packaging should be bio-degradable. Material used for packaging shall not contaminate the food.

8.7.7 Labelling

When the full standard requirements are met, the product can be sold as "Organic". On proper certification by certification agency "India Organic" logo can also be used on the product.

8.7.8 Storage and Transport

Products integrity should be maintained during storage and transportation of organic products. Organic products must be protected from co-mingling with non-organic products and must be protected all times from contact with the materials and substances not permitted for use in organic farming.

8.8 CONSTRAINTS OF ORGANIC CERTIFICATION

- Very high fees for certification by certifying agencies.
- Lack of awareness of group certification among small & medium farmers.
- Small and medium-sized Indian organic producers are not able to access international markets because of non certified produce.
- Need of certification in domestic market.
- Under current government policy and as per the standards, it takes approximately three years for a farm in conversion to be certified as organic, and costs are hefty for the small farmer.
- While internationally, out exports cannot hold ground without certification, uncertified brands in India have had some success. Still the domestic market has not developed for the certified organic produce and demand for safe, pure and healthy food is slowly increasing.
- In developed countries, especially in Europe and the US, every supermarket has a green line where a complete range of certified organic products is available. India is just beginning to manage to hold in some supermarkets.

Chapter 9

Processing and Labelling of Organic Produce

Processing and handling organic products should be optimized to maintain the quality and integrity of the product. The operator must develop an organic production and handling plan. The important points to be considered in handling the organic produce are discussed as under:

9.1 SPECIFIC REQUIREMENTS

(i) Description of practices and procedures to be performed.

(ii) Description of the record keeping system implemented to comply with the requirements of NPOP.

(iii) List of each substance/input used during production, storage and handling indicating its composition, source, location where it will be used and documentation of commercial availability as applicable.

(iv) Description of the monitoring practices and procedures followed and maintained to verify the plan is effectively implemented.

(v) Organic products shall be protected from co-mingling with non-organic products during parallel processing and handling.

(vi) Pollution sources shall be identified and contamination avoided.

(vii) Processing and handling of organic products should be done separately in time or place from handling and processing of non-organic products.

(viii) All products shall be adequately identified through the whole process.

(ix) Certification programme shall regulate the means and measures to be allowed /recommended for decontamination, cleaning or disinfection of all facilities where organic products are kept, handled, processed or stored.

9.2 INSECT PESTS AND DISEASE CONTROL

(i) Pests and diseases should be avoided by following good agricultural practices. This includes general cleanliness and hygiene.

(ii) Treatments with pest regulating agents must thus be regarded as the last resort.

(iii) Recommended treatments are physical barriers, sound, ultra-sound, light and UV-light, traps (incl. pheromone traps and static bait traps, temperature control, controlled atmosphere and diatomaceous earth).

(iv) A plan for pest prevention and pest control should be developed.

(v) For pest management and control the following measures shall be used in order of priority:

- Preventive methods such as disruption, elimination of habitat and access to facilities
- Mechanical, physical and biological methods
- Pesticidal substances permitted under national standards
- Other substances used in traps

(vi) Irradiation is prohibited.

(vii) There shall never be direct or indirect contact between organic products and prohibited substances. (e.g. pesticides). In case of doubt, it shall be ensured that no residues are present in the organic product.

(viii) Persistent or carcinogenic pesticides and disinfectants are not permitted.

9.3 INGREDIENTS, ADDITIVES AND PROCESSING AIDS

(i) 100 per cent of the ingredients used in processing shall be organic except where an organic ingredient is not available in sufficient quality or quantity, non-organic ingredients may be used to a minimum extent only in case of essential technological need or for particular nutritional purpose. Such non-organic raw material shall not be genetically engineered. The accredited Certification Body may authorize the use of non-organic raw materials subject to periodic re-evaluation.

(ii) The same ingredient within one product shall not be derived both from an organic and non organic origin.

(iii) Preparations of micro-organisms and enzymes commonly used in food processing may be used, with the exception of genetically engineered micro-organisms and their products. For the production of enzymes and other micro-biological products, the medium shall be composed of organic ingredients.

(iv) Water and salt may be used in organic products.

(v) Minerals (including trace elements), vitamins and similar isolated ingredients shall not be used. The certification programme may grant exceptions where use is legally required or where severe dietary or nutritional deficiency can be demonstrated.

(vi) Ethylene gas is permitted for ripening.

9.4 PROCESSING METHODS

(i) Processing methods should be based on mechanical, physical and biological processes.

(ii) The vital quality of an organic ingredient shall be maintained throughout each step of its processing methods and shall be chosen to limit the number and quantity of additives and processing aids. The following kinds of processes are approved :

- Mechanical and physical
- Biological
- Smoking
- Extraction
- Precipitation
- Filtration

(iii) Extraction shall be either with water, ethanol, plant and animal oils, vinegar, carbon dioxide, nitrogen or carboxylic acids. These shall be of food grade quality, appropriate for the purpose.

(iv) Filtration substances shall not be made of asbestos nor may they be permitted with substances which may negatively affect the product.

(v) Irradiation is not allowed.

9.5 PACKAGING

(i) Eco-friendly biodegradable, recyclable, reusable packaging materials shall be used wherever possible.

(ii) Packaging materials that affect organic nature of the content should be avoided.

(iii) Use of PVC material is prohibited. Laminates and aluminium should be avoided.

(iv) The packages shall be closed in such a manner that substitution of the content cannot be achieved without manipulation or damage of the seal.

(v) The accredited Certification Body shall approve the packaging material for use.

9.6 LABELLING

9.6.1 Labelling Requirements

(i) Labelling shall convey clear and accurate information on the organic status of the product.

(ii) When the full standards requirements are fulfilled, products shall be sold as "produce of organic agriculture" or a similar description.

(iii) The label for conversion products shall be clearly distinguishable from the label for organic products by mentioning the year of conversion.

(iv) The name and address of the person or company legally responsible for the production or processing of the product shall be mentioned on the label.

(v) Product labels should list processing procedures, which influence the product properties in a way not immediately obvious. All components of additives and processing aids shall be declared.

(vi) Additional product information shall be made available on request.

(vii) Ingredients or products derived from wild production shall be declared as such.

9.6.2 Processed Products

(i) Single ingredient products may be labelled as "Organic" when all standard requirements have been met.

(ii) Multi ingredient products where not all ingredients, including additives, are of organic origin may be labelled in the following way (raw material weight):

- Where a minimum of 95 per cent of the ingredients are of certified organic origin, products may be labelled "certified organic" or similar and should carry the logo of the certification programme.
- Where less than 95 per cent but not less than 70 per cent of the ingredients are of certified organic origin, products may not be called "organic". The word "organic" may be used on the principal display in statements like "made with organic ingredients" provided there is a clear statement of the proportion of the organic ingredients. An indication that the product is covered by the certification programme should be used, close to the indication of proportion of organic ingredients.
- Where less than 70 per cent of the ingredients are of certified organic origin, the indication that an ingredient is organic may appear in the ingredients list. Such product may not be called "organic".

(iii) Added water and salt shall not be included in the percentage calculations of organic ingredients. For aquaculture products the use of iodized salt shall be referred on the labels.

(iv) All raw materials of a multi-ingredient product shall be listed on the product label in order of their weight percentage. It shall be apparent which raw materials are of organic certified origin and which are not. All additives shall be listed with their full name.

(v) If herbs and/or spices constitute less than 2 per cent of the total weight of the product, they may be listed as "spices" or "herbs" without stating the percentage.

(vi) Organic products shall not be labelled as GE (genetic engineering) or GM (genetic modified) free in order to avoid potentially misleading claims about the end product. Any reference to genetic engineering on product labels shall be limited to the production method.

(vii) The label of a certified organic product must depict the name and logo of the accredited Certification Body, accreditation number and India Organic Logo.

(viii) The accredited Certification Body shall verify the labelling requirement and approve the labels of their certified operators before the labels are used.

9.7 STORAGE AND TRANSPORT

(i) Organic products shall be stored at ambient temperature. The following special conditions of storage are permitted

- Controlled atmosphere
- Cooling
- Freezing
- Drying
- Humidity regulation

(ii) Product integrity should be maintained during storage and transportation of organic products. Organic products must be protected at all times from co-mingling with non-organic products and from contact with materials and substances not permitted for use in organic farming and handling.

(iii) Where only part of the unit is certified and other products are non-organic, the organic products should be stored and handled separately to maintain their identity.

(iv) Bulk stores for organic product should be separated from conventional product stores and clearly labelled to that effect.

(v) Storage areas and transport containers for organic product should be cleaned using methods and materials permitted in organic production. Measures should be taken to prevent possible contamination from any pesticide or other treatment.

The detailed description of various regulations governing organic production can be obtained from APEDA or downloaded from their website. These stringent production requirements have been laid down to help to produce the food commodities suitable for export. Organic food market is growing worldwide at a fast pace.

Therefore, development of production practices in accordance with regional conditions and farmers requirement is need of the hour. India can utilize economical manpower to produce cost competitive and high quality products for export and for domestic market. Organic production is one area of agriculture which can convert India's **'Green Revolution'** into **'Evergreen Revolution'**. However, it will require highest level of commitment from every individual and institution engaged directly or indirectly in promotion of organic farming. Who knows we might be looking at another billion dollar industry in the making similar to information technology.

Chapter 10

Economic Viability of Organic Farming

Organic farming is known to be environmentally sustainable, but it be economically sustainable as well. During last few years, farmers have shown steadily increasing interest in organic farming. Many farmers who adopted organic farming methods early in this period were motivated by reasons relating to the health and safety of their families, consumers, and livestock, and by idealistic convictions about soil and land stewardship. More recently, as cost of chemicals and credit have increased and commodity prices have stagnated, thousands of conventional farmers have begun to search for ways to decrease input costs. "Low input farming" or "Zero budget farming" are the new, socially acceptable identical terms for organic farming, and economic survival is the motivation for many new comers.

Prices for organic foods reflect many of the same costs as conventional foods in terms of growing, harvesting, transportation and storage. Organically produced foods must meet strict regulations governing all these steps so the process is often more labour and management intensive, and farming tends to be on a smaller scale. There is also mounting evidence that if all the indirect costs of conventional food production (cleanup of polluted water, replacement of eroded soils, costs of health care for farmers and their workers) were factored into the price of food, organic foods would cost the same, or, more likely be cheaper.

10.1 ECONOMIC VIABILITY

In general crop yields under organic management are somewhat lower than conventional system. Yield reduction of 10–20 per cent in some lands and similar or higher in others. Studies have shown that organic agriculture is economically viable, that farmers can achieve more income as a result of premiums vis-a-vis they need fewer inputs to maintain higher net returns. Organic systems are based on the optimum use of local resources and technologies and can give farmers greater independence and more control over their means of production. However, more comprehensive monitoring is needed to analyze their sustainability and impact.

Many farmers enter organic production because they want to maintain their farm in a more holistic way. Other major stimuli for developing organic agriculture include environmental and social concerns, economic necessity, a lack of chemical inputs, and market demands. Small farmers are also encouraged to take up and stay in organic farming by the prospect of being able to produce more food at the subsistence level, having a larger surplus for local sale, or being able to cultivate a product of significant export value. Farmers are most responsive to organic agriculture when they have not been exposed to the chemical message and their farming systems involve traditional or negligible inputs from market. When production is relatively labour intensive and if farmers have the chance of developing the organic concept themselves they are also more inclined to convert to organic agriculture. However, farmers are less likely to take up organic farming in situations of high labour cost and labour scarcity and where there has been an over exposure to the chemical farming. Farmers with relatively mechanized farms and a commitment to high input, high output strategies are also less likely to convert.

Organic farming is economically viable because:

- Reduction in the use of external inputs and increase in on-farm inputs with the greatest potential to benefit the health of farmers and consumers.
- More productivity through the incorporation of natural processes as nutrient cycles, nitrogen fixation, and pest-predator relationships into the agriculture production process.
- Greater productive use of the biological and genetic potential of plant and animal species.
- Improvement of the match between cropping patterns

and the productive potential and physical limitations of agricultural lands to ensure long term sustainability of current production levels and

- Profitable and efficient production with emphasis on improved management and conservation of soil, water, energy biological resources.

10.2 Economics of Organic Farming

The replacement of external inputs by farm – derived resources normally leads to a reduction in variable input costs under organic managements. However, in a few cases, a higher input costs in organic farming is due to the purchase of compost and other organic manures and bio-pesticides from market have been reported. Studies have shown that the common organic agricultural combination of lower input costs and favorable price premiums can offset reduced yields and make organic farms equally or often more profitable than conventional farms.

The interest in organic agriculture in developing countries is growing because it places more reliance on the natural and human resources available, requires less financial input and provides safe food while conserving the environment. Studies to date seem to indicate that organic agriculture offers comparative advantage in areas with less rainfall and relatively low natural and soil fertility levels. Labour realizes a good return and this is important where paid labour is almost non-existent. Organic agriculture does not need costly investments in irrigation, energy and external inputs, but rather organic agricultural policies have the potential to improve local food security, especially in marginal areas. Thus through greater emphasis on use of local resources and self-reliance, conversion to organic agriculture definitely contributes to the empowerment of farmers and local communities.

10.3 Economics of Different Cropping Systems

The studies conducted under Network project on Organic Farming revealed that across the locations, net return was 17 per cent higher (at 20-25 % premium price) under organic production system compared to inorganic production system. The cost of cultivation was found to be 13 per cent higher under organic production system mainly due to handing of bulky nature of organic manures. Benefit: Cost ratio of important cropping systems experimented under National Project on Organic Farming (NPOF) is given in table 10.1.

Table 10.1. Economics of Different Cropping Systems in Organic Farming.

Cropping System (s)	B : C ratio*	Cropping System (s)	B : C ratio*
Babycorn -Potato-Greengram	2.38	Rice -Lentil-Sesbania (GM)	0.97
Brinjal-Sunflower	1.41	Rice -Mustard-Sesbania (GM)	1.19
Cabbage-Radish-Capsicum	0.81	Rice -Pea (veg.)Sesbania(GM)	2.16
Cauliflower-Radish-Tomato	1.42	Rice-Dolichos bean	0.94
French bean-Cauliflower-French bean	0.86	Rice-Groundnut	1.35
Ginger	1.97	Rice-Maize	1.12
Groundnut -Sorghum	3.52	Rice-Mustard	0.75
Maize-Cotton	2.60	Rice-Pea-Sorghum (F)	2.99
Maize-Garlic	1.83	Rice-Potato-Radish	1.91
Maize-Mustard-Radish-Greengram	1.87	Rice-Wheat-Sesbania (GM)	1.35
Maize-Potato-Okra	1.73	Sorghum (F)-Pea-Okra	2.69
Potato-Chickpea	3.06	Soybean- Wheat	3.00
Rice – Berseem	3.46	Soybean-Berseem	1.58
Rice - Carrot	4.34	Soybean-Chickpea	2.11
Rice – French bean	3.18	Soybean-Isabgol	2.44
Rice -Mustard	1.04	Soybean-Mustard	1.94
Rice - Potato	2.06	Turmeric	1.91
Rice – Potato -Okra	3.34	Turmeric+ Onion	1.26
Rice –Barley + Mustard-Greengram	1.90		

* Benefit cost ratio worked out by taking 20 to 25 % premium price for organic products & it is mean of over the years and across the locations.

* GM – Green manure, F– Fodder, veg. – vegetable

Chapter 11

Marketing and Export Potential of Organic Products

In last few years very rapid growth in organic farming has took place particularly in Australia, Argentina, USA, Spain, China, Italy, France, Uruguay, India and Germany. The ill effects of chemicals used in agriculture have changed the mindset of some consumers in different countries who are now prefer to buy organic food even with high premium for health. Organic farming is also promoted for restoration of soil health and creating better environment. Presently organic farming is being practiced in 181 countries of the world. The global organic area is 69.8 mha (2017). Total area under organic certification process in India is 3.43 mha. This includes 1.94 mha (65.5%) cultivable area and another 1.49 mha (43.5%) for wild harvest collection (APEDA, 2019). The world organic market is now 97.0 billion USD (FIBL & IFOAM, 2018). The organic market in India is valued at INR 51. 5 billion (APEDA, 2019) with annual growth rate of 12-15 per cent.

Export is driving organic food production in India. The increasing demand for organic food products in the developed countries and the extensive support by the Indian government coupled with its focus on agri- exports are the drivers for organic food industry. As the per capita income of the most of Indian population is quite low, hesitate to purchase organic food at 20-30 per cent premium price, therefore domestic market is not sufficient to consume the entire organic food produced in the country. As a result, export of organic food is the prime aim of organic farmers as well as of the government. Agriculture Processed Food Export Developed Authority (APEDA) coordinates the export of organic food and other food products in India. Since organic standards developed by

NPOP considering the international organic production standards such as CODEX and IFOAM, Indian organic food products are readily accepted in the US and European markets.

11.1 GLOBAL STATUS OF ORGANIC FOOD MARKET

At global level, the market for organic food is expanding vigorously. According to an Ecovia Intelligence and FiBL survey, the data on the organic retail sale of food and beverages showed a Compound Annual Growth Rate (CAGR) of 10.4 per cent from 2000 to 2016 with the global sale having increased to USD 90 billion in 2016 from USD17.9 billion in 2000. Further the, global organic food and beverages market was valued at approximately USD 124.76 billion in 2017 and is expected to generate revenue of around USD 323.09 billion by the end of 2024, growing at a CAGR of around 14.56 per cent between 2017 and 2024 (ZION Market Research, 2018). Organic Food and Beverages Market by Product Type Organic Food (Organic Fruits & Vegetables, Organic Meat, Fish & Poultry, Organic Dairy Products, Organic Frozen & Processed Foods and Others) and Organic Beverages (Organic Non-Dairy Products, Organic Coffee & Tea, Organic Beer & Wine and Others). North America is expected to dominate the market with the highest CAGR during the forecast period. North America held approximately 40 per cent of the global market revenue share. Europe accounts for over 33 per cent of total revenue share globally. Germany is expected to lead market in Europe whereas various other countries like UK, France, Spain, Russia, and Italy are expected to witness rapid growth in organic food and beverages in the region.

Asia Pacific is anticipated to exhibit a rapid increase in CAGR during the forecast period, owing to the changing lifestyles and increase in consumer disposable income. The Asian market has a significant amount of imports of processed organic food and beverages from the developed countries of Europe and North America. The industrialized and developed countries such as Australia, New Zealand, Japan, Singapore, Hong Kong, and Korea are major markets for organic foods and beverages in the Asia Pacific. There is also an increase in demand for the products in developing countries like India and China due to awareness about the benefits of organic food.

Some of the key players in the global organic food and beverages market include The Hain Celestial Group Inc., Amy's Kitchen Inc., Organic Valley, Private Label Companies, Whole Foods Market Inc., General Mills Inc., Everest, Cargill Inc., White Wave Foods, and United Natural Foods Incorporated, etc. Major players are frequently engaged in research and development of new product range with enhanced shelf life and its nutrients.

11.2 INDIAN STATUS IN GLOBAL ORGANIC FOOD MARKET

India produced 2.56 mt (2018-19) of certified organic products which includes all varieties of food products namely oilseeds, sugarcane, cereals & millets, cotton, pulses, medicinal plants, tea, fruits, spices, dry fruits, vegetables, coffee, etc. The total volume of export during 2018-19 was 0.614 mt. The organic food realization was around INR 51.51 billion (757.49 million USD). In terms of export value realization, oilcake/meal (23.35%) lead the products followed by oilseeds (16.40%), processed foods (16.20), cereals and millets (8.38%), plantation crops (7.93%), spices & condiments (5.70%) and dry fruits (5.33%). The organic food export has risen by almost 160 per cent between 2015-16 and 2018-19 from INR 19.76 billion to INR 51.51 billion. Though it is a very small portion of Indian overall agricultural export basket (less than 3%), organic food has shown a constant increase in the last few years. In fact, in the past decade organic food exports have grown from around INR 5.0 billion a year to almost 51 billion. Europe union, Netherlands, United Kingdom, Germany, Belgium, Sweden, Switzerland, France, Italy, Spain, Americas, USA, Canada, Middle East, Saudi Arabia, UAE, Asia, Japan, Singapore, Australia, Africa, South Africa are the major destinations for India's organic food exports, which have lot of potentials.

Indian organic food market is projected to grow at a CAGR of over 23 per cent by 2023, on account of government policies supporting organic farming coupled with rising land area under organic cultivation.

Some of the major players operating in India organic market include Sum Inter India Organics Private Limited, Nature Bio-Foods Limited, Organic India Pvt. Ltd, Sresta Natural Bio-products Pvt. Ltd., Morarka Organic Food Pvt. Ltd., etc.

Organic Food Products Exported From India

- Organic cereals : Wheat, rice, maize
- Organic pulses : Redgram
- Organic fruits : Mango, orange, pineapple, passion fruit, casheunut, walnut
- Organic oilseeds and oils – Soybean, sunflower, mustard, cotton seed, groundnut, castor
- Organic vegetables – Garlic, onion, brinjal, potato, tomato
- Organic herbs & spices – Chilli, cardamom, peppermint, turmeric, black pepper, white pepper, amla, tamarind, ginger, clove, vanilla, cinnamon, nutmeg
- Others – Jaggery, sugar, tea, coffee, cotton, textiles.

11.3 INDIAN DOMESTIC MARKET AND EXPORT

Although India has initiated export market for organic foods successfully, the situation in relation to domestic organic market is quite miserable. No doubt the organic food consumption in India is on the rise but at a very slow pace. Some people believe that organic food is only a "concept" popular in the developed countries. They think that when it comes to organic food, India only exports organic food and very little is consumed. However, this is not true. Though 50 per cent of the organic food production in India is targeted towards exports, there are many who look towards organic food for domestic consumption. The most important reason for buying organic food is the concern for the health of children, with over 66 per cent parents preferring organic food to non-organic food. Though organic food is priced over 25 per cent more than conventional food in India, many parents are willing to pay this higher premium due to the perceived health benefits of organic food. The increase in organic food consumption in India is evident from the fact that many organic food stores are spurring up in India. Today every supermarket has an organic food store and every large city in India has numerous organic food stores and restaurants. This is a huge change considering that the first organic food store in Mumbai was started in 1997. Organic food exports from India are increasing with more farmers shifting to organic farming. With the domestic consumption being low, the prime market for Indian organic food industry lies in the United States and Europe. India has now become a leading supplier of organic herbs, organic spices, organic basmati rice, etc. The increasing demand for organic food products in the developed countries and the extensive support by the Indian government coupled with its focus on agri-exports are the drivers for the Indian organic food industry. Organic food products in India are priced about 20-30 per cent higher than non-organic food products. This is a very high premium for most of the Indian population where the per capita annual income is merely INR 1,45,000 (USD 2041) at present. The domestic market is not sufficient to consume the entire organic food produced in the country. As a result, exports of organic food is the prime aim of organic farmers as well as the government.

The domestic market for organic products is yet not developed as that of export market. The products available in the domestic market in organic quality are rice, wheat, cotton, tea, coffee, pulses, fruits and vegetables. Wholesalers/traders and super markets play major roles in the distribution of organic products. As most organic production originates from small farmers, wholesalers / traders account for a 60 per cent share in the distribution of organic products. Large organized producers distribute their products through super markets as well as through self-owned stalls. Considering the profile of existing consumers of organic products, super markets and restaurants are the major marketing channels for organic products. Major markets for organic products lie in metropolitan cities – Mumbai, Delhi, Kolkata, Chennai, Bangalore, Hyderabad and other cities.

11.3.1 Limitations in Growth of Domestic Market

- Farmers do not know where to sale and consumers do not know from where to get organic food. Market is limited and supply is inconsistent.
- Apathy of farmers to grow organically as sale is not ensured because demand is low and expected premium is not received as the response from consumer is poor.
- Farmers cannot identify as to what crop will be grown organically, what price will be charged, etc.
- Lack of consumer awareness on the benefit of organic produce.
- The quality supply is not ensured. The cost towards certification is very high. Farmers are not aware about the agencies for certification and analysis.
- Logistic of slow shipment, restriction for importing Indian organic products.
- Time consuming and complicated paper work while dealing with export authorities.
- Less effort to develop domestic market.

11.3.2 Steps to Create Domestic Market

- Development of crop- specific/location-specific organic package of practices for different crops and cropping systems.
- Development of rural market where the arrangement for farmers to consumers direct selling can be made with a premium price.
- Registration of consumers for supply of organic foods and development of home delivery system.
- Ease the certification procedure and make it cost effective for the farmers.
- Govt. should also make all efforts to promote organic farming.
- Appropriate training to the farmers and NGOs to create awareness about benefits of organic farming/ products.
- Awareness program for consumers regarding benefits of organic produce.

11.4 CONSTRAINTS AND OPPORTUNITIES

(i) The demand for organic products has created new export opportunities for the developing world. While some consumers express a preference for locally-grown organic foods, the demand for a variety of foods year-round makes it impossible for any country to source organic food entirely within its own borders. As a result, many developing countries have begun to export organic products successfully. Typically, organic exports are sold at impressive premiums, often at prices 20 per cent higher than identical products produced on non-organic farms. The ultimate profitability of organic farm varies, however, few studies have a assessed the long-term potential for such market premiums. Nevertheless, under the right circumstances the market returns from organic agriculture can potentially contribute to local food security by increasing family incomes.

(ii) Farmers are denied access to developed countries organic markets for two to three years after beginning organic management since such countries will not certify land and livestock as organic before that time, arguing that it is necessary for the purging of chemical residues. However, products produced on land under organic management for at least one year but less than the two-three years standard can be sold as *transitional organic,*although few markets have developed for such products.

(iii) In most cases farmers and post-harvest businesses seeking to sell their products in developed countries must hire an organic certification organization to annually inspect and confirm that these farms and businesses adhere to the organic standards established by various trading partners. The cost for this service can be expensive, although it varies in relation to farm size, volume of production, and the efficiency of the certification organization (e.g. IFOAM certification costs a maximum of 5 per cent of sales value). Few developing countries have certification organization within their borders, and even when sufficient resources are available to pay for certification famers often lack the information to find credible inspectors.

(iv) While most developing countries have focused on export markets in the developed world domestic market opportunities for organic food or eco-food may also be exploited.

(v) Whether the intent is to sell organic products domestically or abroad, reliable market information is difficult to obtain. There is virtually no systematic production or market survey data being collected with which to assess the rate and pattern of organic market growth. In particular no projections for the market in the developing world have been made nor have markets systematically been identified for developing country exports. Estimates of the public's willingness to pay premium of the impact of regional and tastes and the incidence of markets have not been undertaken.

(vi) There is a growing consumer demand in the developed countries for healthy food. This change is the consumer preference reflects the increasing concern about the dangers then associated with the consumption of food containing small quantities of food additives and pesticide residues. As the organically produced food is devoid of such harmful additives and residues, it is regarded as healthy food and as such, enjoys special consumer preference. To the average consumer in the developed countries, organically produced food is more acceptable then the food produced under the conventional agriculture system because of the special qualities possess by the former in regard to test and nutrition.

(vii) In the developed countries organically produced food has not yet gained consumer acceptance to any appreciable extent. This is despite the presence of harmful chemical residues at higher concentrations in vegetables, fruits and other farm produce in India, 40 times more than the amount what an average American or European consumes. These pesticide residues lead to all kinds of chornic diseases of the heart, brain, kidney and liver including cancer.

(viii) There is lack of awareness among the people of the developing countries about poisons in the food they are consuming daily. Even for those who are aware of the dangers of poison in food, they do not have easy access to organic products. Wherever such products are available they are costlier than the conventionally produced items mainly because organic agriculture does not enjoy the benefit of subsidy. However, there are still parts of India and other developing countries where people have access only to organic (chemical free) foods because agrochemicals have not yet used.

(ix) On the other hand in the developed countries there is a rapid awareness with regard to organically grown and processed foods and related issues. In Europe and USA in addition to Reform houses, Departmental Stores is exclusively for organic products. In the west super markets are generally considered as a threat by environmentalists, but they may still have a role to play. By introducing a wholly different segments of the consumer population to organic foods super markets can help to increase strength of organic products, improve the consistency of supply by creating greater demand and increase overall customer awareness of organic issues. This super market can plough the vast masses of consumers turning non users into light user and light users into regular customers of organic especiality stores. Problem for especiality stores and super markets is consumer resistance to the higher price of organic products. Inspite of this, consumer who are aware and willing to pay higher price for such products not only because the taste is better, but also because of the realization that in the process they are supporting organic farming and the thus protecting the environment.

(x) Organic agriculture is not only leading the way to viable alternatives to modern development model, but it is also part of the solution to global environment problems. Now, more than ever opportunities exist for the organic market to extend from less than 1.0 per cent of overall food cells to between 10 and 20 per cent by the return of the century. There are, however, critical and sometimes very sensitive issues to be addressed in order to sustain a growing market for colgate learning organically product produced food.

11.5 MARKET CHALLENGES

- The rampant use of pesticides and chemicals for insect pests and weed control, has caused an evolution of the pest and weed species immune to chemicals.
- As organic farming prohibits synthetic pesticides, the vulnerable farmer is at the mercy of severe attacks from mutant pests.
- Severe lack of suitable infrastructure is proving to be the Achilles' heel for organic farming and its produce.
- Poor road infrastructure especially in the hilly states results in poor and delayed connectivity to farmer markets.
- Inadequate agricultural infrastructure and cold storage facilities translate to loss of produce due to spoilage.
- An acute shortage of effective and viable methods of organic pest control.

11.6 RECOMMENDATIONS TO PROMOTE DOMESTIC AND EXPORT MARKET

- Government should devise a mechanism to cover losses to farmers due to reduced yields by combination of lowering input costs and favorable price premium for the organic produce.
- Advertising campaigns advocating the health benefits to consumers are imperative to promote organic produce in domestic market in order to create substantial demand.
- Government should make labeling compulsory for all the products available in market for sale. The specifications required on label must be standardized to rule out chances of any confusion regarding certification (Participatory Guarantee System or third Party).
- State Agricultural Universities should develop region specific package of practices for organic crop rotation with focus on IPM and bio-pesticides.
- Further, the same package of practices (PoPs) need to be transferred to farmers through extensive extension activities. In this regard, government should earmark funds for

research and development to develop package of practices and extension activities.

- Stringent measures are required to check the sale of spurious bio-pesticides and bio-fertilizers in the market to maintain the organic integrity.
- It is suggested to design the training programmes and implement the same to impart required skills to the organic farmers through KVKs.
- Farmers must also be sensitized about the certification process to facilitate certification.
- The infrastructure needs to be created to promote organic agriculture like organic fertilizers and organic pesticides units. Such units should be established at village level where farmers are practicing organic farming.
- Strong market support with proper infrastructure should be created for, like adequate road connectivity, refrigeration container in trains, kisan mandis for organic produce, etc. for expansion of organic agriculture in the country.
- A stringent mechanism needs to be devised to ensure that the higher price paid by consumer for organic produce should reach the farmer and is not simply absorbed by intermediaries in the chain.
- Regulations for sale of organic products certified through accredited certification bodies should be notified under the Food Safety and Standards Act.

Glossary

Agricultural-ecosystem/agro ecosystem: It is composed of the total complex of the crops/animals in an area together with overall environment and as modified by management practices.

Agro-ecology: The study of the relation of agricultural crops and environment.

Agroforestry: A collective term for systems of land use in which woody plants (trees and shrubs) are deliberately combined on the same land management unit with herbaceous crops and or animals; either in some form of spatial arrangement or in sequence. For a land use system to fall within the concept of agroforestry, there should be both ecological and economic interactions between woody plants and other components of the system. It may involve the integration of trees into farming systems of crops and livestock into forests.

Agronomy: The word has been derived from Greek Word "agros" meaning fields and "nomos" meaning management. It is a branch of agriculture dealing with crop production and soil management.

Amritpani: Mix 10 kg cow dung with 500 g honey and mix thoroughly to form a creamy paste. Add 250 g of cow *desi ghee* and mix at high speed. Dilute with 200 litre water. Sprinkle this suspension in one acre over soil or with irrigation water. After 30 days apply second dose in between the row of plants or through irrigation water.

Beejamrutha: Beejamrutha, a mix off of cow dung, cow urine, water, lime and a handful of soil, a totally organic product helpful for the plant growth and protects crop from harmful and soil borne and seed borne pathogens.

Biodiversity: A measure of the variety of species comprising a community.

Biodynamic: A type of organic farming system developed by the scientist and philosopher Rudolf Steiner in the early 1600's. Biodynamic farming takes into consideration both biological cycles and also "dynamic"-metaphysical or spiritual - aspects of the farm, with the intention of achieving balance between physical and non-physical realms.

Bio-fertilizers: Bio-fertilizers or microbial fertilizers or more appropriately microbial inoculants' are preparations containing live or latent cells of efficient strain of nitrogen fixing micro-organisms used for seed or soil application with the objective of increasing the number of such micro-organisms in soil or rhizosphere and consequently improve the extent of micro-biologically fixed nitrogen for plant growth. They are used either to fix nitrogen or to solubilise plant nutrients like phosphates or to otherwise stimulate plant growth through synthesis of growth promoting substances or to collect available phosphorus from remote places out of reach of plant root hairs by sending elongated filaments (VAM).

Bio-intensive: A combination of biodynamic and the French - intensive method of farming, which involves the use of raised beds, with crops planted very close together and in combination with other crops.

Biological control: The practice of using beneficial organisms – such as insect predators or parasites of pest insects, pest diseases agents, insect-eating birds and bats to keep pest populations at a tolerable level.

Biological methods: The successful control of pests and diseases by means of another organisms that is encouraged and disseminated by man is known as bio-control. This means use of any biotic agents for minimizing the pest population either directly or indirectly.

Bulky organic manures: They contain small percentage of nutrients and applied in large quantities. Farm yard manure, compost, vermi-compost, crop residues and green manure are most important bulky organic manures. They have several advantages as (i) supply all essential plant nutrients (ii) improve physical conditions of soil (soil structure, water holding capacity, aeration, buffers soil surface temperature), (iii) increases availability of nutrients, (iv) stimulates soil flora and fauna, and (v) control parasitic nematodes and fungi to some extent by altering balance of micro-organisms in the soil.

Catch crop: (i) A crop grows temporarily on land that will eventually carry a crop of different species, in order to maintain the productivity or prevent the erosion of the land prior to the establishment of the main crop (ii) A crop employed to stimulate germination and growth of some specific, parasitic weed and then destroyed before planting the main crop.

Certified farmers' market: A marketplace in which farmers sell their produce directly to consumers. In a certified farmer's market farmers are exempt from packing, sizing, and labelling requirements, however, they can only sell products that they have produced.

Certified organic: Referring to a product that has been produced in accordance with specific regulations and that has been inspected and approved by an accredited certifying agency. The USDA Federal Rule governing organic certification requires that an organic production system is managed "to respond to site-specific conditions by integrating cultural, biological and mechanical practices that foster cycling of resources, promote ecological balance, and conserve biodiversity."

Compost: Compost is a mass of rotted organic matter made from decomposition of plant and animal wastes with various additives. The compost made from the farm wastes like weeds, paddy straw, sugarcane trash and other plants and wastes is called **farm compost.**

Composting: (i) A biological process in which micro-organisms of both types, namely, aerobic and anaerobic, decompose the organic matter and lower the carbon nitrogen ratio of the refuse. The final product of composting is a well rotted manure known as compost, **(ii)** Composting is the natural process of 'rotting' or decomposition of organic matter by micro-organisms under controlled conditions. Raw organic materials such as crop residues, animal wastes, food garbage.

Concentrated organic manures: Concentrated organic manures are those materials that are organic in nature and contain higher percentage of essential plant nutrients such as nitrogen, phosphorus and potash, as compared to bulky organic manures. These concentrated manures are made from raw materials of animals or plant origin. The concentrated organic manures commonly used are aquatic weeds, oilcakes, blood meal, fish meal, meat meal and horn & hoof meal.

Conversion requirements: The time between the start of organic management and cultivation of crops or animal husbandry is known as the conversion period. All standard requirements should be met during conversion period. Full conversion period is not required where organic farming practices are already in use.

Conversion: Turning a non-organic agricultural holding into an organic one. The lengthy and complicated process is required by regulations in order to be able to use organic logos and labelling.

Cover crop: **(i)** A crop grown to prevent soil erosion by covering the soil with living vegetation and roots that hold on to the soil. Cover crops are also grown to help maintain soil organic matter and increase nitrogen availability (green manure crop), and to "hold on" to excess nutrients (a catch crop) still in the soil following an economic crop. Other benefits of cover crops include weed suppression and attraction of beneficial insects **(ii)** A close growing crop grown primarily for the purpose of protecting and improving soil between periods of regular crop production or between trees and vines in orchards and vineyards. Cover crops are known to reduce the danger of soil erosion and to protect the land from leaching losses.

Critical period: The shortest time span in the ontogeny of plant growth when a treatment will result in maximum effect.

Crop ecosystems: Cropping systems ranging between monoculture and multispecies culture of fields and garden crops, singly or in combination and their relation to environmental conditions and management practices.

Crop residue: After crop harvesting, the left over plant material including leaves, stalk and roots is known as crop residue.

Crop rotation: The practice of planting a sequence of different crops and cover crops on a specific field. Crop rotations can be used to help build soil fertility, reduce insect, pest pressure, and suppress weeds.

Duration of conversion period: The establishment of an organic management system and building of soil fertility requires an interim period, the conversion period. The **conversion period** may not always be of sufficient duration to improve soil fertility and reestablish the balance of the ecosystem but it is the period in which all the actions required to reach these goals are started.

Eco-conscious: Someone who is referred to as eco-conscious is conscious of the environment and the ecosystems that support it. People who are eco-conscious are aware of how our buying habits, living habits and eating habits impact the earth.

Eco-farming: Potential for introducing mutually reinforcing ecological approaches to food production.

Ecology: Study of the interrelationship between living organisms with each other and with their environment.

Ecosystem: An ecosystem has been defined as an open system gases, water and minerals that are involved together in the flow of energy and circulation of matter. An ecosystem results from the integration of all living and non-living factors of the environment for a defined segment of space and time.

Fairomones: These are volatile compounds that evoke behavioral response adaptively favourable to the receiver. Fairomones are released either by host plant or by the host insects while former issued by the pest and natural enemies to locate their habitats, the later is used for prey finding and parasitisation preying.

Farmyard manure (FYM): Farmyard manure is the traditional organic manure. The term 'FYM' refers to decomposed mixture of dung and urine of farm animals along with litter and left over material from roughages or fodder fed to the cattle. FYM contains 0.5 % N, 0.25 % P_2O_5 and 0.5 % K_2O, on an average basis.

Genetically modified organism (GMO): An organism that has been genetically altered through the transfer of DNA from another organism, resulting in the expression of new characteristics in the recipient.

Good agriculture practice (GAP): It is a collection of principles, processes, resulting in safe and healthy food and non food agricultural products, while taking into account economical, social and environment sustainability.

Green manure: A cover crop grown to help in maintaining soil organic matter and increase nitrogen availability. Legumes are often used because they have rhizobial bacteria living in their root nodules that are able to fix nitrogen from the air and add it to the soil. Grasses grow quickly, providing biomass good for increasing organic matter.

Green manuring: Green manuring can be defined as a practice of ploughing or turning into the soil under composed green plant tissues for improving physical condition of soil as well as soil fertility. From the time immemorial, the turning in a green crop for improvement of the soil conditions has been a popular farming practice. Green-manuring, wherever feasible, is the principal supplementary means of adding organic matter to the soil. It consists of planting of a fast growing crop and ploughing it under to incorporate it into the soil. The green manure crop supplies organic matter as well additional nitrogen, particularly if it is a legume crop, which has the ability to fix nitrogen from the air with the help of its root nodule bacteria. The green-manure crops also exercise a protective action against erosion and leaching.

Harvest index: The ratio of grain yield to biological yield. It is a factor less than unity, say 0 to 05.55, but some workers prefer to use harvest index (%) in which they express the index as 0 to 55 per cent. It is expressed as; harvest index = grain yield/ biological yield or total yield. The determination of harvest is usually based on a harvest at maturity.

Homa organic farming: Homa organic farming is holistic healing for agriculture and can be used in conjunction with any good organic farming system. It is obviously extremely inexpensive and simple to undertake but requires discipline and regularity.

Intercropping: The practice of planting two or more mutually beneficial crops in close proximity, typically as alternating rows or number of rows (on a small scale, this often called companion planting). Benefits can include weed suppression, structural support, or shade.

Intercrops: The crop raised in an orchard or other widely spaced crops for increasing the income from the same piece of land e.g. short duration vegetables, pulses, oilseeds, etc.

Interculture: Operations to be carried out after sowing or planting in between crop rows includes operations like gap filling, thinning, weeding, mulching and top dressing. May be done by manual labour or by bullock or machine drawn implements.

Intergraded pest management (IPM): A strategy of pest management that focuses on long term prevention of pests of their damage through a combination of techniques such as biological control, habitat manipulation, modification of cultural practices, and use of resistant varieties. Pesticides are used only after monitoring, indicates they are needed according to established guidelines, and treatments are made with the goal of removing only the target organism. Pest control materials are selected and applied in manner that minimizes risks to human health, beneficial and non-target organisms, and the environment.

Intergrated fertility management: Technologies for management of fertilizing the crops with the objective to produce optimum crop yields with minimum fertilizer inputs taking into consideration ecological and socio-economic resources available for supply to the crop under a given agro-ecosystem, without deteriorating the soil-fertility, such as application of organic manures, green manures, blue green algae and bio-fertilizers along with use of inorganic fertilizers for achieving higher yields and also improving the soil productivity.

Jiwamrita: It is an organic fertilizer which enriches the soil and plant and provides all the nutrients required for the growth of the plant. It is a rich source of nitrogen and valuable micro-organisms which naturally enhances soil fertility.

Ley farming: System of growing grass in rotation with agriculture crops; extensively used in U.K. and U.S.A.; helps to improve soil fertility and providers good fodder to animals; also help in build up of soil structure and prevent soil erosion. Can be conveniently adopted in places like Nilgiris abounding in good pasture land and having suitable climate.

Mixed farming: Farming system which involves the raising of crop and rearing of animals and / or poultry. Mixed farming is based upon the principle that land should support animal and animal should support land.

Mulching: The practice of spreading organic materials such as straw, compost, or wood chips – over otherwise bare soil between and among crop plants. Mulching helps to conserve moisture, suppress weeds, and build soil organic matter.

Myco-herbicide: A biological agent for weed control involving import of a pathogen from an area of co-evolution with its host and release it into a new area where the host has become a weedy host in the absence of its parthogen. It involves the use of micro-organisms as herbicide by applying them, to target weeds in a manner similar to herbicides.

Mycorrhiza: The morphological association, usually symbiotic fungi with the roots of seed plants. The association is referred to as ectotrophic, the fungal hyphae occur on the root surface and penetrate only the intercellular spaces and as endotrophic when the hyphae occur mainly within the cells of the host plant.

Natural farming: It is not alternative system of farming but part of the philosophy of life involving continuous search to know the true spirit and form of nature.

Non-symbiotic nitrogen fixation: Fixation of molecular nitrogen by free living, micro organisms like clostridium and Azotobactor.

Organic certification: Organic certification is a certification process for producers of organic food and other organic agricultural products. In general, any business directly involved in food production can be certified, including seed suppliers, farmers, food processors, retailers and restaurants. Requirements vary from country to country, and generally involve a set of production standards for growing, storage, processing, packaging and shipping that include.

Organic farming: United States Department of Agriculture (USDA) study team on organic farming defined organic agriculture as "organic farming is a system which avoids or largely excludes the use of synthetic inputs (such as fertilizers, pesticides, growth regulators, livestock feed additives, etc.) to the maximum extent feasible, rely upon crop rotations, crop residues, animal manures, legumes, green manure, off-farm organic wastes, mechanical cultivation, mineral bearing rocks, and aspects of biological pest control to maintain soil productivity and tilth, to supply plant nutrients, and to control insects, weeds, and other pests". It considers that soil is a living system which must be "fed" in a way that does not restrict the activities of beneficial organisms necessary for recycling of nutrients and producing humus at centre to this concept.

Organic matter: Any material that was recently living or produced by a living organism and is capable of being decomposed.

Organic nitrogen (soil): Nitrogen in the form of organic compounds e.g. amino acids, amino sugar, purine and pyramidine bases.

Organic phosphorus: Phosphorus present as a constituent of an organic compound or a group compounds, e.g. glycerophosphoric acid, inosital phosphoric acid, cytidylic acid.

Organic production: A production system that is managed accordance with the act and regulations and respond to site-specific conditions by integrating cultural, biological, and mechanical practices that foster cycling of resources, promote ecological balance, and conserve biodiversity.

Organic: A labeling term that refers to an agricultural product produced in accordance with the act and the regulations in this part.

Panchgavya krishi: *Panchgavya* is a special bio-enhancer prepared from five products obtained from cow dung, urine, milk, curd and *ghee*. Dr Natrajan, a Medical practitioner and scientist from Tamil Nadu Agricultural University, Coimbtore has further refined the formulation suiting to the requirement of various horticultural and agricultural crops.

Panchgavya: Mix fresh cow dung 5 kg, cow urine 3 litre, cow milk 2 litre, curd 2 litre, cow butter oil 1 kg and ferment for 7 days with twice stirring per day. Dilute 3 litre of panchgavya in 100 litre water and spray over soil. Twenty litre of panchgavya is needed per acre for soil application along with irrigation water.

Permaculture: A term coined in 1978 by Bill Mollsion, Australian ecologist, and one of his students, David Holmgren. "Permaculture" stands for "permanent agriculture' and it is a land use concept that refers to the design of ecological human habitats and food production systems, with goal of harmonious integration of human dwellings, annual and perennial plants, animals, soil, and water, into stable, productive communities.

Permanent crop land area: Land cultivated with crops that need not be replanted after each harvest, such as rubber, coffee, cocoa, including land under shrubs, fruit trees, nut trees and vines but excluding land under trees grown for food or timber.

Pheromone traps: Pheromone is ectohormons secreted by an organism, which elicit behavioral responses from other members/ sex of its own species. These are extremely selective, non-toxic, highly bio-degradable and effective at low application rates. Synthetic sex pheromones are commercially available and are used for surveillance, monitoring and control of many *Lepidopterous* pests such as spotted bollworm, tobacco catterpiller, potato tuber moth, diamond back moth and leaf folder, etc.

Pollution: Contamination of natural environment by the addition to air or water or substance potentially toxic or otherwise harmful to man and animals for example, SO_2, Co, insecticides, etc.

Rishi krishi: Drawn from Vedas, the Rishi Krishi method of natural farming has been mastered by farmers of Maharashtra and Madhya Pradesh. In this method, all on-farm sources of nutrients including composts, cattle dung manure, green leaf manure and crop biomass for mulching are exploited to their best potential with continuous soil enrichment through the use of Rishi Krishi formulation known as " *amritpani*" and virgin soil.

Sanjivak: Mix 100 kg cow dung, 100 litre cow urine and 500 g jaggary in 300 litre of water in a 500 litre closed drum. Ferment for 10 days. Dilute with 20 times water and sprinkle in one acre either as soil spray or along with irrigation water.

Soil organic matter (SOM): Soil organic matter has three parts: living organisms, fresh residues, and well decomposed residues (the living, the dead, and the very dead). Fresh residues are a primary source of food for living organisms. Decomposition of fresh residues releases nutrients needed by plants. Well decomposed matter, also called " humus" holds on to some to nutrients, storing them for slow release to plants.

Soil solarization: It is the act of using the solar energy of the sun as an environmentally friendly way to kill weeds and control soil-borne pathogens such as fungi, nematodes, insects, or bacteria.

Stale seed bed: A seed bed left unsown to let a heavy flush of weeds appear and be destroyed with tillage operation or by herbicide seeding the crop.

Sustainable agriculture: Referring to an agriculture system that is ecologically sound, economically viable, and socially just– a system capable of maintaining productivity indefinitely.

Symbiotic nitrogen fixation: Bacteria belonging to genus *Rhizobium* are capable of fixing atmospheric nitrogen in association with legume crops. *Rhizobium* spp. enter the roots of host plants and form nodule on the root surface. The bacteria depend on the host plant for carbohydrates and water. *Rhizobium* supplies nitrogen to the host. Nitrogen fixed by the *Rhizobium* is translocated through xylem vessels of the host plant mainly in the form of aspergine and to some extent as glutamine. The colour of nodule varies from whitish to brown, green and pink. The pink coloured nodules due to the presence of pigment legheamoglobin are called effective nodules.

Transitional: Referring to a production systems which follows organic management practices, but has not yet fulfilled time requirements to be certified organic (land must be free from prohibited materials for a minimum of three years to be certified).

Trap crop: A crop that is planted to lure pest insects away from an economic crop.

Vermicast: A single worm casting or a quantity of worm castings. Worms 'work' material by ingesting, excreting, and re-ingesting it. Vermicast is extensively worm worked and re-worked. It may be overworked and has probably lost plant nutrients as compared vermicompost. Vermicast has a fine, smooth texture, which may dry with a crust on the surface.

Vermicompost: Mixture of partially decomposed organic waste, beddings, and worm castings (excretions). Contains recognizable fragments of plant, food, or worm bedding material, as well as cocoons, worms, and associated organisms. As a verb, " to carry out composting with worms".

Vermiculture: The technique of breeding and raising earthworms in controlled conditions scientifically is called **vermiculture,** and making compost with the use of earthworms is called vermicomposting. About 1000 adult earthworms can convert 5 kg waste into compost per day.

Vermiwash (a liquid manure): Vermiwash a liquid collected after the passage of water through a column of worm action is very useful as a foliar spray to enhance the plant growth and yield and to check development of diseases. It is a collection of excretory products and mucus secretions of earthworm along with nutrients from the soil organic molecules.

Weed control methods: Weed control in crop fields is the process of limiting any given weed infestation to the extent that it permits economic crop production.

Weed Prevention: Weed prevention comprises all measures which deny the entry and establishment of weeds in an area. In other words, all practices that help to discourage the weeds from becoming a problem over time, for the subject of weed prevention.

Appendix

APPENDIX – I

Acronyms

APEDA	Agricultural and Processed Food Product Export and Development Authority, New Delhi
IFOAM	International Federation of Organic Agriculture Movement Bonn, Germany
USDA	United States Department of Agriculture
INM	Integrated Nutrient Management
IPM	Integrated Pest Management
IDM	Intergrated Disease Management
IWM	Intergrated Weed Management
FIBL	Research Institute of Organic Agriculture, Switzerland
NCOF	National Centre of Organic Farming, Ghaziabad
NPOP	National Programme for Organic Production
BD	Biodynamic
PKVY	Paramparagat Krishi Vikas Yojna
NMSA	National Mission for Sustainable Agriculture
SHM	Soil Health Management
RKVY	Rashtriya Krishi Vikas Yojna
NPOF	National Project of Organic Farming
NHM	National Horticulture Mission
NFSM	National Food Security Mission
LEIA	Low External Input Agriculture
ITK	Indigenous Technical Knowledge
FYM	Farm Yard Manure

PSM	Phosphate Solubilizing Microorganisms
VAM	Vesicular Arbuscular Mycorrhiza
GV	Granulosis Virus
NPV	Nuclear Polyhedrosis Virus
CFU	Colony Farming Unit
BT	Bacillus Thuringiensis
DAS	Days After Sowing
IWN	Integrated Weed Management
NOP	National Organic Program
JAS	Japanese Agricultural Standard
NAPP	National Accreditation Policy and Programme
NSOP	National Standards for Organic Production
NAB	National Accreditation Body
IQS	Internal Quality System
FAO	Food & Agriculture Organization
CAGR	Compound Annual Growth Rate

APPENDIX – II

Products for Use in Fertilizing and Soil Conditioning

Matter Produced on an Organic Farm Unit		
a.	Farmyard & poultry manure, slurry, urine	Permitted
b.	Crop residues and green manure	Permitted
c.	Straw and other mulches	Permitted
1. Matter Produced Outside the Organic Farm Unit		
a.	Blood meal, meat meal, bone meal and feather meal without preservatives	Restricted
b.	Compost made from any carbon based residues (animal excrement including poultry)	Restricted
c.	Farmyard manure, slurry, urine	Restricted
d.	Fish and fish products without preservatives	Restricted
e.	Guano	Restricted
f.	Human excrement	Not allowed
g.	By-products from the food and textile industries of restricted bio-degradable material of microbial, plant or animal origin without any synthetic additives	Restricted
h.	Peat without synthetic additives (prohibited for soil conditioning)	Permitted
i.	Sawdust, wood shavings, wood, provided it comes from untreated wood	Permitted
j.	Seaweed and seaweed products obtained by physical processes, extraction with water or aqueous acid and/or alkaline solution	Restricted
k.	Sewage sludge and urban composts from separated sources which are monitored for contamination	Restricted
l.	Straw	Restricted
m.	Vermicasts	Restricted
n.	Animal charcoal	Restricted
o.	Compost and spent mushroom and vermiculate substances	Restricted
p.	Compost from organic household reference	Restricted
q.	Compost from plant residues	Permitted

r.	By products from oil palm, coconut and cocoa (including empty fruit bunch, palm oil mill effluent (pome), cocoa peat and empty cocoa pods)	Restricted
s.	By products of industries processing ingredients from organic agriculture	Restricted
2. Minerals		
a.	Basic slag	Restricted
b.	Calcareous and magnesium rock	Restricted
c.	Calcified seaweed	Permitted
d.	Calcium chloride	Permitted
e.	Calcium carbonate of network origin (chalk, limestone, gypsum and phosphate chalk)	Permitted
f.	Mineral potassium with low chlorine content (e.g. sulphate of potash, kainite, sylvinite, patenkali)	Restricted
g.	Natural phosphates (e.g. Rock phosphates)	Restricted
h.	Pulverised rock	Restricted
i.	Sodium chloride	Permitted
j.	Trace elements (baron, In, Fe, Mn, molybolerum, Zn)	Restricted
k.	Woodash from untreated wood	Restricted
l.	Potassium sulphate	Restricted
m.	Magnesium sulphate (Epson salt)	Permitted
n.	Gypsum (calcium sulphate)	Permitted
o.	Stillage and stillage extract	Permitted
p.	Aluminum calcium phosphate	Restricted
q.	Sulphur	Restricted
r.	Stone mill	Restricted
s.	Clay (bentonite, perlite, zeolite)	Permitted
3. Microbiological Preparations		
a.	Bacterial preparations (bio-fertilizers)	Permitted
b.	Biodynamic preparations	Permitted
c.	Plant preparations and botanical extracts	Permitted
d.	Vermiculate	Permitted
e.	Peat	Permitted

APPENDIX– III

Products for Plant Pest and Disease Control

I. Substances from plant and animal origin		
a.	*Azadirachta indica* [neem preparations (neem oil)]	Restricted
b.	Preparation of rotenone from *Derris elliptica, Lonchocarpus, Thephrosia spp.*	Restricted
c.	Gelatine	Permitted
d.	Propolis	Restricted
e.	Plant based extracts (e.g. neem, garlic, pongamia, etc	Permitted
f.	Preparation on basis of pyrethrins extracted from Chrysanthemum, Cinerariaefolium, containing possibly a synergist pyrethrum cinerafolium	Restricted
g.	Preparation from *Quassia amara*	Restricted
h.	Release of parasite predators of insect pests	Restricted
i.	Preparation from Ryania species	Restricted
j.	Tobacco tea	Not allowed
k.	Lecithin	Restricted
l.	Casein	Permitted
m.	Sea weeds, sea weed meal, sea weed extracts, sea salt and salty water	Restricted
n.	Extract from mushroom (Shiitake fungus)	Permitted
o.	Extract from Chlorella	Permitted
p.	Fermented product from Aspergillus	Restricted
q.	Natural acids (vinegar)	Restricted
II. Minerals		
a.	Chloride of lime soda	Restricted
b.	Clay (e.g. bentonite, perlite, vermiculite, zeolite)	Permitted
c.	Copper salts / inorganic salts (Bordeaux mix, copper hydroxide, Inspection and Certification Agency used as fungicide, maximum 8 kg per ha per year depending upon the crop and under the supervision of Inspection and Certification Agency	Restricted
d.	Mineral powders (stone meal, silicates)	Not allowed

e.	Diatomaceous earth	Restricted
f.	Light mineral oils	Restricted
g.	Permanganate of potash	Restricted
h.	Lime sulphur (calcium polysulphide)	Restricted
i.	Silicates (sodium silicate, quartz)	Restricted
j.	Sodium bicarbonate	Permitted
k.	Sulphur (as a fungicide, acaricide, repellent)	Restricted
III. Microorganisms / Biocontrol agents		
a.	Viral preparations (e.g., Granulosis viruses, Nuclear polyhydrosis, viruses etc.).	Permitted
b.	Fungal preparations (e.g., Trichoderma species etc.)	Permitted
c.	Bacterial preparations (e.g., Bacillus species etc.)	Permitted
d.	Parasites, predators and sterilized insects."	Permitted
IV. *Others*		
a.	Carbon dioxide and nitrogen gas	Restricted
b.	Soft soap (potassium soap)	Permitted
c.	Ethyl alcohol	Not allowed
d.	Homeopathic and Ayurvedic preparations	Permitted
e.	Herbal and biodynamic preparations	Permitted

APPENDIX – IV

List of Accredited Certification Bodies under NPOP

Bureau Veritas Certification India (BVCI) Pvt. Ltd., Marwah Centre, 6th Floor Opp. Ansa Industrial Estate, Krishanlal Marwah Marg Off Saki-Vihar Road Andheri (East), Mumbai-400 072 (Maharashtra)	**ECOCERT India Pvt. Ltd.,** Bungalow No. 21A, Shrirang City, Itkheda, Behind Ryan International School, Paithan Road, Auranagabad
IMO Control Pvt. Ltd. No. 3627, 1st Floor, 7th Cross, 13th 'G' Main, H.A.L. 2nd Stage, Bangalore-560 008.	**Indian Organic Certification Agency (INDOCERT)** Kuttamassery, Thottumugham P.O. Aluva-683 105, Cochin (Kerala)
Lacon Quality Certification Pvt. Ltd., Thiruvalla (Kerala) Chenathra, Theepany, Thiruvalla - 689 101 (Kerala)	**OneCert Asia Agri Certification (P) Ltd.,** H-08, Mansarovar Industrial Area, Mansarovar Jaipur-302020, Rajasthan
SGS India Pvt. Ltd. 226,Udyog Vihar, Phase-I Gurgaon-122016 Haryana	**Control Union Certifications,** Mumbai Plot No. C-113, Pawane MIDC, Navi Mumbai - 400709
Uttarakhand State Organic Certification Agency (USOCA) Third Floor, Krishak Bhavan Mussoorie By Pass Ring Road Nehru Gram, Dehradun, Uttarakhand	**APOF Organic Certification Agency (AOCA)** 126, 1st Floor, Govindappa Road, Off D.V.G. Road, Gandhi Bazar, Bangalore-560 004 (Karnataka)
Rajasthan Organic Certification Agency (ROCA) 3rd Floor, Pant Krishi Bhawan, Janpath, Jaipur 302 005 (Rajasthan)	**Vedic Organic Certification Agency** Plot No. 55, Ushodaya Enclave, Mythrinagar, Miyanagar, Hyderabad – 500 050

ISCOP (Indian Society for Certification of Organic Products) Rasi building, 162/163, Ponnaiyarajapuram Coimbatore – 641 001 Tamil Nadu	**Food Cert India Pvt. Ltd** First Floor, Plot No. 69 House No. 8-2-248/1/7/69/1, Beside Cuppa Restaurant Sree Nagarjuna Co-operative Society, Punjagutta Hyderabad 500 082 Andra Pradesh
TQ Certi Services Limited (formerly FoodCert India Private Limited 4th floor, Mithona Towers – I 1-7-80 to 87 Prenderhast Road, Secunderbad Telangana - 500003	**Aditi Organic Certifications Pvt. Ltd** No. 531/A, Priya Chambers Dr. Rajkumar Road, Rajajinagar, 1st Block, Bangalore - 560010
Chhattisgarh Certification Society, India (CGCERT), Raipur A-25, VIP Estate, Khamhardih, Shankar Nagar, Raipur-492007, Chhattisgarh	**Tamil Nadu Organic Certification Department (TNOCD),** Coimbatore 1424 A, Thadagam Road, G.C.T Post, Coimbatore - 641013, Tamil Nadu
Intertek India Pvt. Ltd. E-20, Block B-1, Mohan Cooperative Industrial Estate, Mathura Road, New Delhi - 110 044	**Madhya Pradesh State Organic Certification Agency** Vasundhara, B-II Office Complex Gautam Nagar Bhopal 462 023 Madhya Pradesh
Export Inspection Agency (EIA) Thakkar Bapa Smarak Sadan, 2nd Floor, Dr. Ambedkar Marg (Link Road). Near Jhandewalan Metro Station, New Delhi - 110055	**Odisha State Organic Certification Agency (OSOCA)** Plot No.-326, Baramunda, Bhubaneswar, Odisha, PIN- 751003
Natural Organic Certification Agro Pvt. Ltd. Flat No.302, Aryapratham Apartment, Above Federal Bank, Warje Malwadi, Pune – 411058, Maharashtra	**Fair Cert Certification Services Pvt. Ltd.** C-122, Gauirdham Colony khargone, Madhya Pradesh 451001
Gujarat Organic Products Certification Agency (GOPCA) Beej Pramanan Bhavan Opp. Gokul Row House, Nr. Shayaml Cross Satellite, Ahmadabad 380015, Gujarat	**Uttar Pradesh State Organic Certification Agency** Government Garden Campus Kariayppa Road, Alambagh Lucknow 226005, Uttar Pradesh

Karnataka state Organic Certification Agency Opp. Baptist Hospital Bellary Road, Hebbal Bangalore, Karnataka - 560024	**Sikkim State Organic Certification Agency (SSOCA)** Ground Floor, Soil testing Lab Building, ICAR Complex, Tadong, Gangtok, Sikkim-737102
Global Certification Society Kesar Bagh Colony, Mohal Nihang (Tika Nihang), near Dr. Chaudhary Hospital, Palampur, Himachal Pradesh-176061	**GreenCert Biosolutions Pvt. Ltd.** Office 2, Building No. 12, Runwal Meadow, Warje, Pune-411052
Telangana State Organic Certification Authority HACA Bhawan, 1st Floor 5-10-193, Opp. Public Garden Hyderabad-500004	

APPENDIX – V

References Cited

Bhattacharyya, P. 2004. Organic Food Production in India – Status, Strategy and Scope. Agrobios (India), Jodhpur – 342002

Dhama, A.K. 1999. Organic Farming. Agro-botanical Publishers (India), Bikaner– 334003

Gehlot, Dushyent, 2015. Organic Farming: Components and Management. Agrobios (India), Jodhpur -342002

Jain, Vinod Kumar. 2019. Bio- technology in Sustainable & Organic Farming - Scope and Potential. Oxford Book Company, Jaipur - 302018

Joshi, Mukund. 2016. New Vistas of Organic Farming. Scientific Publishers (India), Jodhpur -342001

Lampin, N. 1990. Organic Farming – Theory and Practice. Scientific Publishers (India), Jodhpur –342001.

Maliwal, P.L. and Mundra S.L. 2018. A text book of weed management. Agrotech Publishing Academy, Udaipur-313002.

Maliwal, P.L. and Mundra, S.L. 2016. Agronomy at a glance (vol. 1) IV Edition. Agrotech Publishing Academy, Udaipur-313002.

Palaniappan, S.P and Annadurai, K. 2018. Organic Farming Theory and Practice. Scientific Publishers (India) Jodhpur - 342001.

Panda, S.C. 2013. Principles and Practices of Organic Farming. Agrobios (India), Jodhpur - 342002.

Purohit, S.S. and Gehlot, Dushyent. 2006. Trends in Organic Farming. Agrobios (India), Jodhpur-342002

Rana, S.S. 2016. Organic Farming. Department of Agronomy, College of Agriculture, Himachal Pradesh, Krishi Vishav Vidyalaya, Palampur – 176062.

Sharma, Arun. K., 2002. A Hand Book of Organic Farming, Agrobios (India) Jodhpur- 342002.

Subba Rao, M.S. 2002. Soil Microbiology. Oxford & IBH, New Delhi.

Thapa, U and Tripathy, P. 2018. Organic Farming in India. Agrotech Publishing Academy, Udaipur – 313002

Veeresh, G.K., Shivshankar, K.C. and Suigbachar, M.A. 1997. Organic Farming and Sustainable Agriculture. Associations for Promoting of Organic Farming, Banglore.

Yadav, A.K. 2010. Organic Agriculture (Concept, Scenario, Principles and Practices). National Centre of Organic Farming, Deptt. of Agriculture and Cooperation, Govt. of India, Ghaziabad, 201001.

Yawalkar, K.S., Agarwal, J.P. and Bokde, S. 1996. Manures and Fertilizers. Agri-horticultural Publishing House, Nagpur 440010.

Related Books

ISBN: 9789389061536

ISBN: 9789388812870

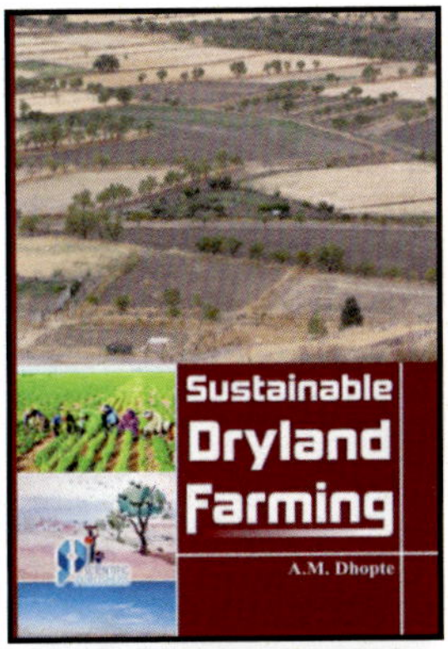

ISBN: 9789388449908

ISBN: 9789388449373

ISBN: 9789387893832

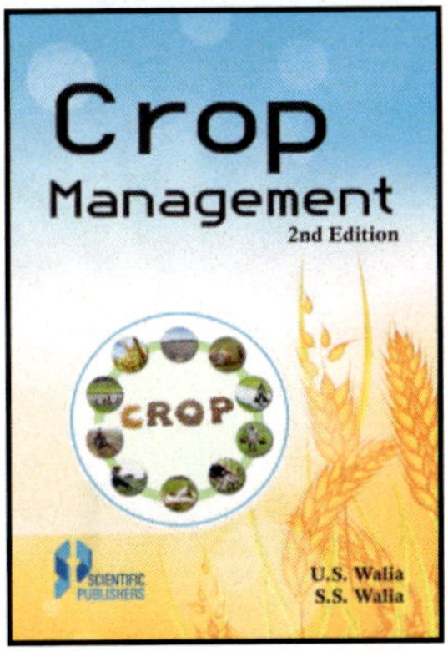

ISBN: 9789388449007

ISBN: 9788172330828

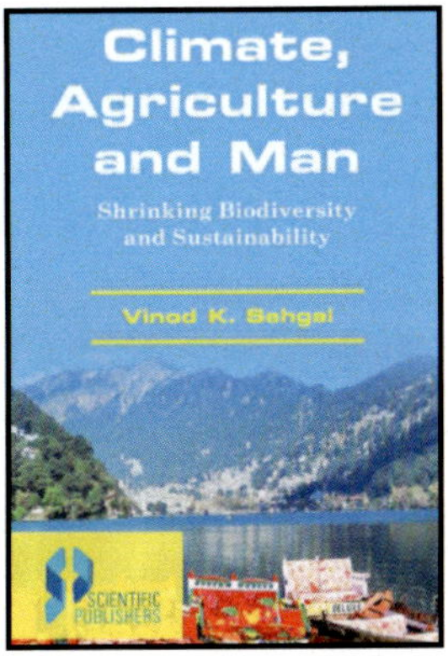

ISBN: 9789387893887

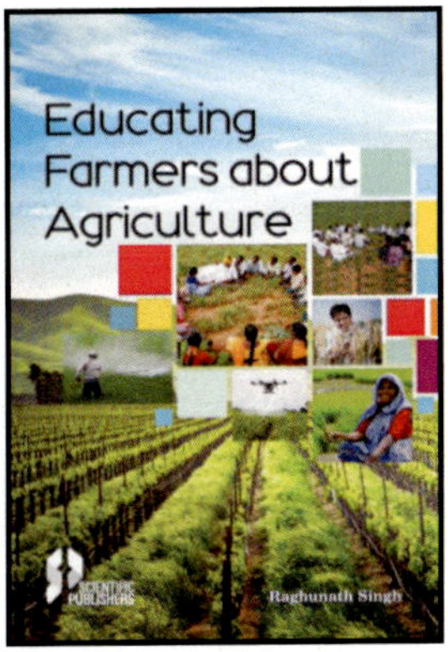

ISBN: 9789388449243

For Full Content & Sample Page Visit Website

To know your local book store call or us on +91-95211-31111

The PR Builders

(arranged in alphabetical order by second name)

Kiran Ray Chaudhury

Madhvendra Das

Tarun Deo

Nikhil Dey

Deepshikha Dharmaraj

Dhrubajyoti Gayan

Veena Gidwani

Santanu Gogoi

Ameer Ismail

Deepak Jolly

Pranav Kumar

Sonya Madeira

Sunayna Malik

Nitin Mantri

Anurag Mittal

Abhilasha Padhy

Pooja Pathak

Udit Pathak

Valerie Pinto

Xavier Prabhu

Sharif Rangnekar

Rekha Rao

Amitabh Saksena

Radhika Shapoorjee

Atul Sharma

Ashwani Singla

Aseem Sood

Rakesh Thukral

Sarvesh Tiwari

Dilip Yadav

LIGHT A SPARK